SpringerBriefs in Environmental Science

SpringerBriefs in Environmental Science present concise summaries of cutting-edge research and practical applications across a wide spectrum of environmental fields, with fast turnaround time to publication. Featuring compact volumes of 50 to 125 pages, the series covers a range of content from professional to academic. Monographs of new material are considered for the SpringerBriefs in Environmental Science series.

Typical topics might include: a timely report of state-of-the-art analytical techniques, a bridge between new research results, as published in journal articles and a contextual literature review, a snapshot of a hot or emerging topic, an in-depth case study or technical example, a presentation of core concepts that students must understand in order to make independent contributions, best practices or protocols to be followed, a series of short case studies/debates highlighting a specific angle.

SpringerBriefs in Environmental Science allow authors to present their ideas and readers to absorb them with minimal time investment. Both solicited and unsolicited manuscripts are considered for publication.

More information about this series at http://www.springer.com/series/8868

Tongbin Chen · Mei Lei · Xiaoming Wan ·
Xiaoyong Zhou · Jun Yang

Phytoremediation of Arsenic Contaminated Sites in China

Theory and Practice

 Springer

Tongbin Chen
Institute of Geographic Sciences and Natural
Resources Research
Chinese Academy of Sciences
Beijing, China

University of Chinese Academy of Sciences
Beijing, China

Xiaoming Wan
Institute of Geographic Sciences and Natural
Resources Research
Chinese Academy of Sciences
Beijing, China

University of Chinese Academy of Sciences
Beijing, China

Jun Yang
Institute of Geographic Sciences and Natural
Resources Research
Chinese Academy of Sciences
Beijing, China

University of Chinese Academy of Sciences
Beijing, China

Mei Lei
Institute of Geographic Sciences and Natural
Resources Research
Chinese Academy of Sciences
Beijing, China

University of Chinese Academy of Sciences
Beijing, China

Xiaoyong Zhou
Institute of Geographic Sciences and Natural
Resources Research
Chinese Academy of Sciences
Beijing, China

University of Chinese Academy of Sciences
Beijing, China

ISSN 2191-5547   ISSN 2191-5555 (electronic)
SpringerBriefs in Environmental Science
ISBN 978-981-15-7819-9   ISBN 978-981-15-7820-5 (eBook)
https://doi.org/10.1007/978-981-15-7820-5

This Springer imprint is published by the registered company Springer Nature Singapore Pte Ltd.
The registered company address is: 152 Beach Road, #21-01/04 Gateway East, Singapore 189721, Singapore

Contents

Chapter 1
Arsenic Hyperaccumulator *Pteris vittata* L. and Its Arsenic Accumulation

Abstract Since its identification as an arsenic (As) hyperaccumulator, the As accumulating and biological characteristics of *Pteris vittata* L. have received extensive investigation. It well satisfies the requirements of an As hyperaccumulator, with the aboveground As concentration higher than 1000 mg kg^{-1}, higher As concentration in shoots than roots, and well growth on heavily As contaminated soil. Furthermore, it has wide distribution in the world, big biomass, extensive root system; and it is able to be harvested more than twice every year. These characteristics make it appropriate plant material for phytoextraction of As from soil. Besides, studies have been conducted on selecting ecotypes able to simultaneously extract multi-metal(loid)s or to be applied to the sustainable management of slightly contaminated soil.

Keywords Arsenic hyperaccumulator · Ecotype · Fern · *Pteris vittata* · Spore

1.1 Arsenic Hyperaccumulating Characteristics of *Pteris vittata*

Phytoremediation, using specific plants to clean up the environment, is a technology that become known to academia in 1990s (Cunningham and Berti 1993; Salt et al. 1998). Phytoremediation, in the broad sense, includes three categories: (1) phytoextraction, using hyperaccumulator, that can accumulate a large amount of toxic contaminants in the aboveground parts, to remove them from soil; (2) phytostabilization, using plants to decrease the bio-availability of contaminants in soil; (3) rhizofiltration, using plants to remove contaminants from water (Salt et al. 1995). Phytoremediation, in the narrow sense, mainly refers to phytoextraction, which is also the main subject of this book. The main targets of phytoextraction are inorganic contaminants, such as arsenic (As), cadmium (Cd), nickel (Ni), and zinc (Zn).

Before the discovery of As hyperaccumulating ability of *P. vittata*, there was no reported arsenic (As) hyperaccumulator. Most know hyperaccumulators were Ni hyperaccumulators. In 2000 and 2001, scientists from China and the US reported the As hyperaccumulation ability of this fern (Chen and Wei 2000; Ma et al. 2001). Through extensive field survey, lab analysis, and greenhouse experiment, it has been

© The Author(s) 2020
T. Chen et al., *Phytoremediation of Arsenic Contaminated Sites in China*,
SpringerBriefs in Environmental Science,
https://doi.org/10.1007/978-981-15-7820-5_1

1

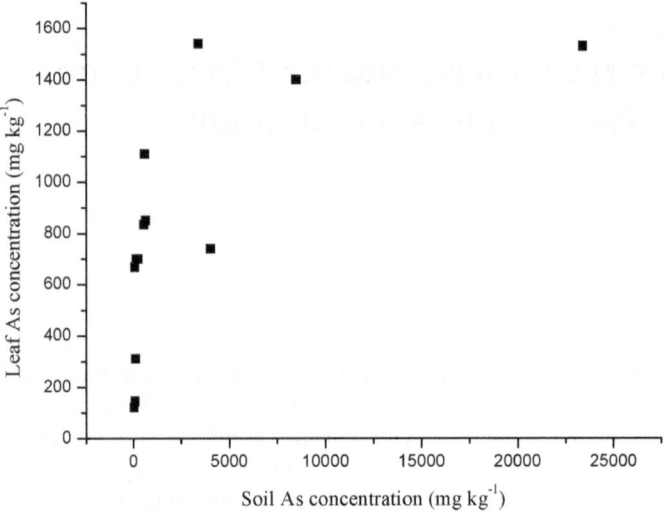

Fig. 1.1 As concentrations in soils and *P. vittata* in the field

confirmed that *P. vittata* is an As hyperaccumulator. It satisfies the criteria of hyperaccumulator: concentrations of As in the aboveground parts higher than 1000 mg kg^{-1}, translocation ratio of As from roots to shoots higher than 1, and no toxics symptoms exposing to high concentrations of As (Chen and Wei 2000).

Field survey indicated that *P. vittata* can normally grow on soil with an As concentration higher than 2% (w:w). The highest As concentration in the leaves (fronds) of *P. vittata* reached 1540 mg kg^{-1}, and the lowest also as high as 120 mg kg^{-1} (Fig. 1.1). This value was thousands of times higher than that in normal plants. And the As concentration in the leaves of *P. vittata* increased with an increase in the concentration of As in soil. But with the As concentration in the soil further increasing from 1% (w:w) to 2.5% (w:w), the As concentration in the leaves of *P. vittata* did not increase.

In the greenhouse experiments (Fig. 1.2), the As hyperaccumulating ability of this fern was further confirmed. The highest As concentration in the leaves (fronds) of *P. vittata* in the greenhouse experiment reached 5070 mg kg^{-1} after six-month growing on a contaminated soil containing 3400 mg As kg^{-1} (Chen et al. 2002). And it is noticeable that the As concentrations in the leaves of *P. vittata* increased linearly with time ($R^2 = 0.999$, $P < 0.01$).

Besides, the translocation ratio of As from roots to shoots was found to be always higher than 1 (Fig. 1.3), despite the provided As concentration in soil. These As hyperaccumulating characteristics indicated high potential of this fern to be applied to phytoextraction of As from contaminated soils.

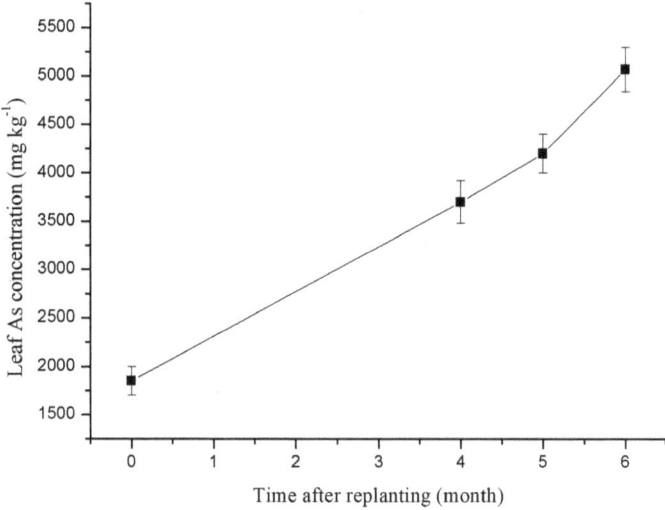

Fig. 1.2 Change in As concentrations *P. vittata* with culture time in the greenhouse

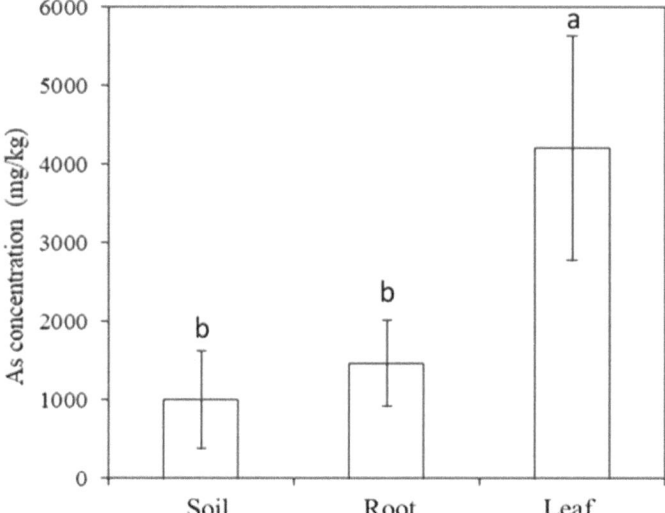

Fig. 1.3 As concentration in soil, root and leaf (frond) of *P. vittata*

1.2 Biological Characteristics of *Pteris vittata*

P. vittata belongs to the genus *Pteris* from *Pteridaceae*. Before the discovery of its As hyperaccumulating ability, it was reported by literature as a Chinese medicine,

Fig. 1.4 Lifecycle of *P. vittata*

able to alleviate rheumatism, hemiplegia, bruises, colds, dysentery, acne, aphids, and snake bites.

P. vittata is a fern species, existing alternately in two forms, namely the diploid sporophyte generation and the haploid protoplast generation (Fig. 1.4). When the sporophyte becomes mature, it will produce haploid spores. Spores germinate on moist soil with warm temperature. The germinated haploid gametophytes (including male antheridia and female archegonia) further fuse into diploid sporophyte. Normally it takes less than one month for spores to germinate, and another one month for sporophyte to form. About 2 months later, sporophyte reached the height of about 5 cm. And when sporophyte reached the height of 15 cm, it is appropriate for the phytoextraction practice in the field. The sporophyte of this fern is a perennial herb, which is 20–200 cm in height, with roots reaching ~30 cm in depth and ~60 cm in diameter (Fig. 1.5).

Other than the reproduction by spores germination, *P. vittata* can also reproduce asexually, including division propagation, and tissue culture. Through division propagation, one mature *P. vittata* plant can be divided into 2–4 plants. While through tissue culture, it normally requires 90–100 days to get *P. vittata* sporelings.

The other characteristic that favors *P. vittata* as a hyperaccumulator is its being perennial. After its harvest, it can quickly sprout new pinnae (Fig. 1.6). In south China, it can be harvested for two to three times every year. The biomass can reach an annual biomass of 36 t (fresh weight) hm^{-2}, enabling the efficient removal of As from soil.

Fig. 1.5   Leaves (left) and roots (right) of *P. vittata*

Fig. 1.6   Harvest (left) and re-sprouting (right) of *P. vittata*

1.3   Distribution of *Pteris vittata* in the World

P. vittata prefers warm, moist and semi-shade environment. It can grow well in full sun and scattered light. According to botanical records, *P. vittata* has a wide distribution throughout the world, ranging from the tropical regions of the East and West Coasts of the North American continent, most islands of the West Indies, rainforest regions of South America, the Mediterranean region of Europe, Northeastern Australia, and Asia, and the tropical and subtropical regions of the ancient continent of North Africa (Fig. 1.7). Recently, it has been spotted in Mexico but not tested for the As hyperaccumulating ability yet.

In China, *P. vittata* is distributed in the vast area south of the Qinling Mountains (Chen et al. 2005). Specifically, *P. vittata* is mainly distributed on alkaline calcareous soils in semitropical or tropical regions.

To investigate the reason for such distribution, and also to get the application range of As phytoextraction technology using *P. vittata*, study was conducted on the effects of temperature and soil pH value on the germination of *P. vittata* spores and the growth of sporophytes. It has been found that such distribution is mainly related

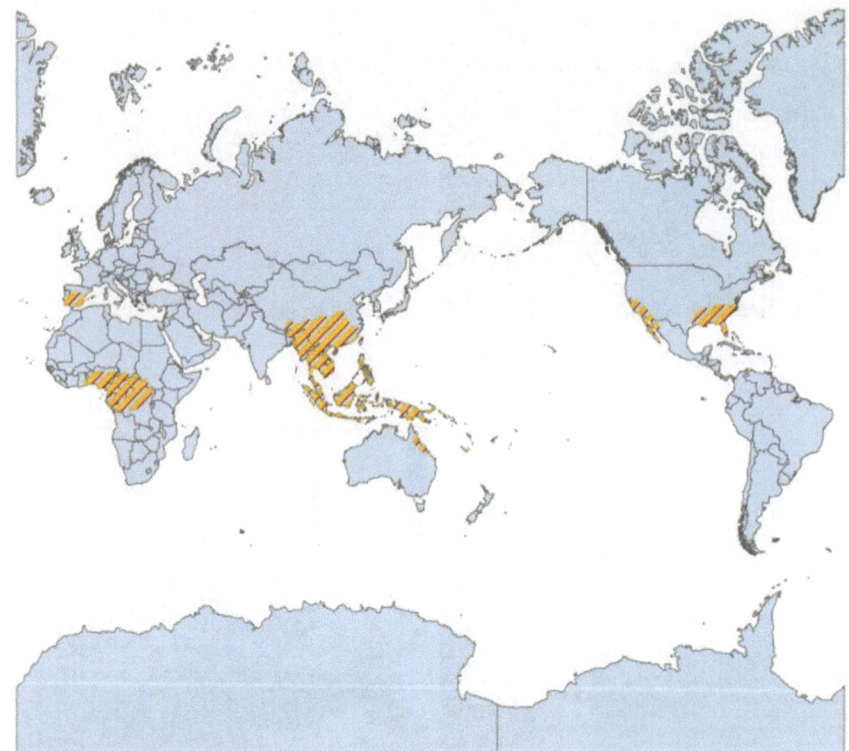

Fig. 1.7 Distribution of *P. vittata* in the world (colored area)

to the environmental conditions required for the spores of *P. vittata* to germinate (Wan et al. 2010).

   With an increase of the pH value of soil, the germination time decreased whereas the germination rate increased, indicating the preference of *P. vittata* to alkaline environment (Fig. 1.8). On soil with the pH value lower than 6, the spores of *P. vittata* cannot germinate. In terms of temperature, 25 °C seems to be the most appropriate, showing the highest germination rate and shortest germination time (Fig. 1.9). Whereas, for the growth of sporophytes afterwards, temperature and soil pH value no longer played a determining role. This indicates that transplantation of *P. vittata* sporelings to areas that it can not naturally germinate might be feasible.

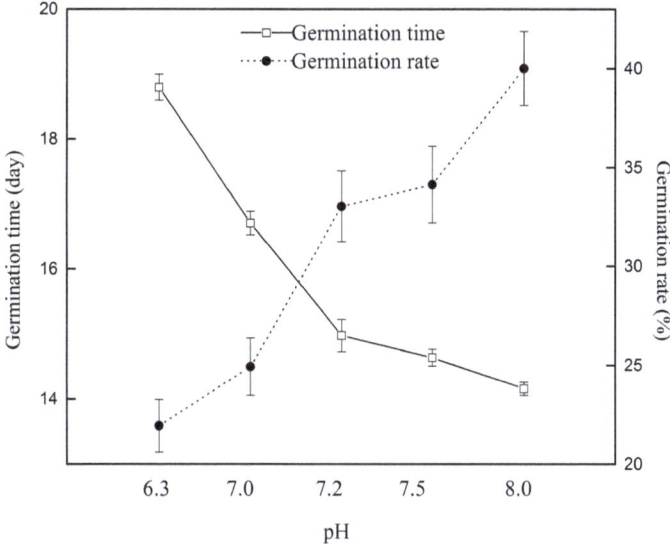

Fig. 1.8 Effects of soil pH on the germination of *P. vittata* spores

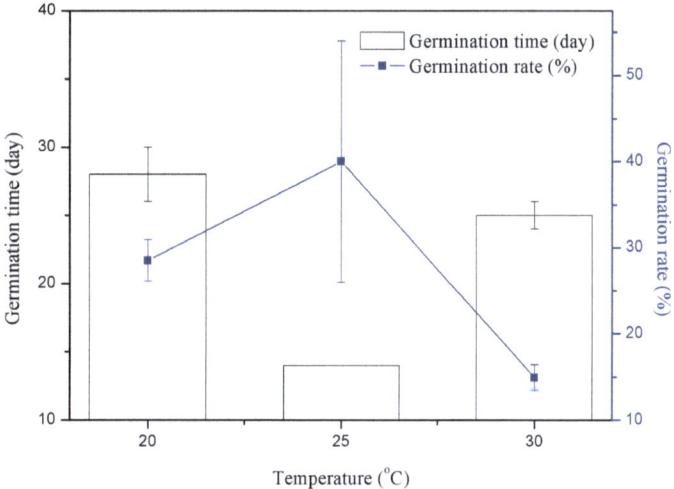

Fig. 1.9 Effect of temperature on the germination of *P. vittata* spores

1.4 Ecotypes of *P. vittata* with Varied as Accumulating Abilities

During the long-term adaptation of *P. vittata* ecotypes to the local environment, different populations of *P. vittata* form varied characteristics that enabled them to

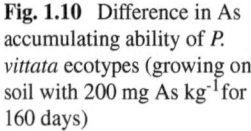

Fig. 1.10 Difference in As accumulating ability of *P. vittata* ecotypes (growing on soil with 200 mg As kg⁻¹ for 160 days)

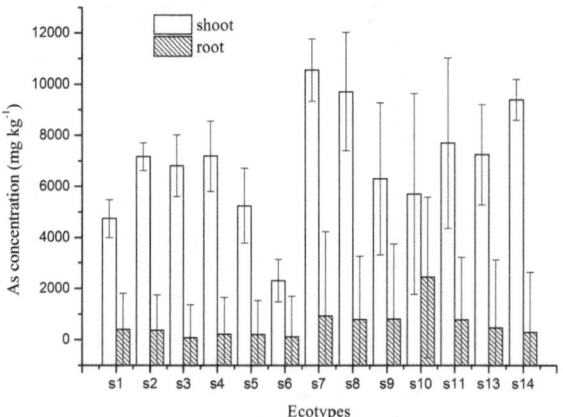

better fit in. It is expected that some characteristics can be used to better remove contaminants from soil. *P. vittata* populations living in different regions and keeping some stable characteristics were categorized into different ecotypes. There are significant differences in As accumulating ability among different ecotypes (Fig. 1.10). Ecotypes are defined as populations with different morphological and physiological characteristics, and these characteristics are genetically stable.

All the identified *P. vittata* ecotypes showed impressive As hyperaccumulating ability: As concentration in leaves (fronds) exceed 2000 mg kg^{-1} after growing on soil with 200 mg As kg $^{-1}$ for 160 days; As concentration in leaf (frond) significantly higher than that in root. Significant difference in As accumulating ability was found among different populations (Wan et al. 2013; Wu et al. 2011). The biggest difference in leaf (frond) As concentration between ecotypes reached five folds.

The habitat As concentration is one of the most important selective forces for the evolution of As hyperaccumulators (Kramer 2010). Four *P. vittata* ecotypes were collected from a mining area with a gradient of soil As concentrations (ranging from 108 to 7527 mg kg^{-1}), to investigate their As accumulation and tolerance after long-term acclimation to difference As stress (Wan et al. 2013). Under field condition, the As concentrations in *P. vittata*, ranging from 402 to 4207 mg kg^{-1}, were proportional to the soil As concentration (Table 1.1). Under greenhouse condition, ecotypes from low As environment had 80% higher As accumulation capacity than that from high As habitat, which may be attributed to the stronger affinity to As of this ecotype's As transporters. In contrast, the ecotype from a high As environment had 5 times higher As tolerance than that from low As habitat, which was related to the more efficient sequestration of As in inactive areas as inactive forms.

Results suggest that As tolerance and As accumulation are two separate processes in hyperaccumulator *P. vittata*, influenced oppositely by habitat As concentration. It was speculated that As tolerance was formed under As stress while As accumulation was more likely a congenital trait of *P. vittata*.

Other than ecotypes with different As accumulating abilities, studies are conducted on ecotypes with multi-metal(loid)s accumulating abilities (Feng et al.

Table 1.1 Field investigation and lab experiments on the four *P. vittata* ecotypes

Ecotype	Field investigation				Lab experiment			
	Soil	*P. vittata*			Soil	*P. vittata*		
	Total As (mg kg^{-1})	As in shoot (mg kg^{-1})	As in root (mg kg^{-1})	TF	Total As (mg kg^{-1})	As in shoot (mg kg^{-1})	As in root (mg kg^{-1})	TF
HN1	108	465^a ± 104	328^a ± 162	1.75^b ± 1.11	200	2436^a ± 90	583^a ± 109	4.3^a ± 0.9
HN2	1159	1003^a ± 624	682^a ± 359	1.52^b ± 1.02	200	1871^b ± 272	760^{ab} ± 292	2.7^b ± 0.8
HN3	3579	402^a ± 109	506^a ± 242	1.02^b ± 0.77	200	1923^b ± 297	858^b ± 146	2.3^b ± 0.2
HN4	7527	4207^b ± 1426	1463^b ± 549	2.94^a ± 0.56	200	1356^c ± 364	753^{ab} ± 75	1.8^b ± 0.5

Different letters within a column indicate significant difference among ecotypes ($P < 0.05$, $n = 3$). TF means translocation factor, the quotient of As in shoot divided by As in root

2016; Huang et al. 2018; Mullner et al. 2013; Tisarum et al. 2014; Wan et al. 2014, 2016, 2017b; Wu et al. 2009, 2013; Xiao et al. 2008) and ecotypes that can better adapt to intercropping mode (Wan and Lei 2018).

It has been found that an ecotype from As and lead (Pb) co-contaminated area has a potential to co-accumulate As and Pb (Fig. 1.11). Due to the long-term evolution process in the habitat with high concentration of Pb, this ecotype can detoxify high concentrations of Pb. And there observed a positive interaction between As and Pb: the addition of As can improve the uptake of Pb, and vice versa.

P. vittata could survive in pot soils spiked with 80 mg kg^{-1} of Cd and tolerated as great as 301 mg kg^{-1} of total Cd and 26.8 mg kg^{-1} of diethyltriaminepenta acetic acid (DTPA)-extractable Cd under field conditions (Xiao et al. 2008). The highest concentration of Cd in fronds was 186 mg kg^{-1} under a total soil concentration of 920 mg As kg^{-1} and 98.6 mg Cd kg^{-1} in the field. The Cd-tolerant ecotype of *P. vittata* extracted effectively As and Cd from the site co-contaminated with Cd and As, and might be used to remediate and revegetate this type of site.

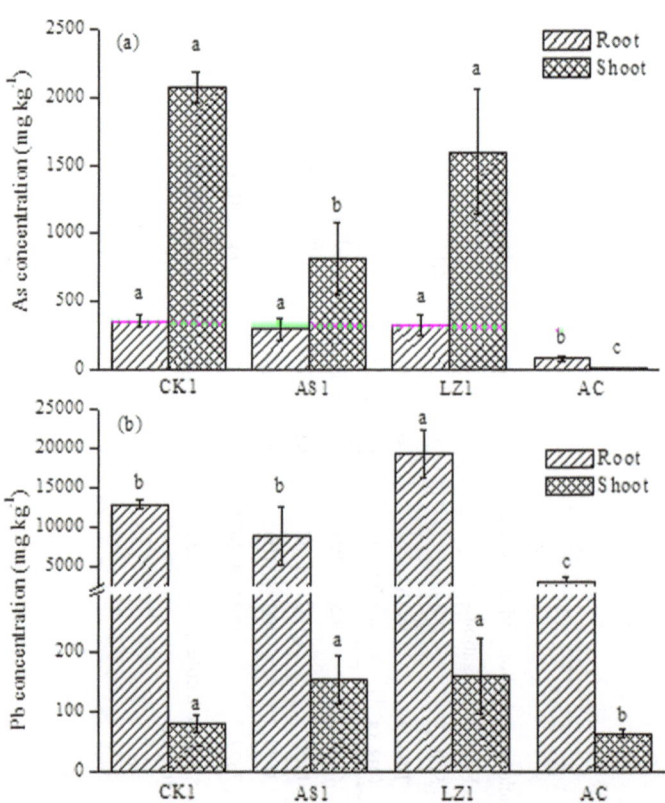

Fig. 1.11 Screening out the As and Pb co-accumulating ecotype of *P. vittata*. LZ1 ecotype was collected from an area with high concentrations of As and Pb concentration, which shows a potential to co-accumulate As and Pb (Wan et al. 2014)

Similarly, the potential of As hyperaccumulator *P. vittata* to remediate sites co-contaminated with Zn and As was investigated (An et al. 2006). *P. vittata* had a very high tolerance to Zn and grew normally at sites with high Zn concentrations. In addition, *P. vittata* could effectively take up Zn into its fronds, with a maximum of 737 mg kg^{-1} under field conditions. The high Zn tolerance, relatively high ability to accumulate Zn, and great capacity to accumulate As under conditions of suppression by high Zn suggest that *P. vittata* could be useful for the remediation of sites co-contaminated with Zn and As.

Intercropping is a multiple cropping practice involving growing two or more crops in proximity to produce a great yield on a given piece of land. Intercropping has been practiced by farmers for more than 2000 years in China, greatly improving the land use efficiency (Zhang and Li 2003). Recently, intercropping has been proposed to be an alternative for the soil remediation, which can clean the soil and simultaneously produce safe agricultural products. The intercropping technology has been well utilized in several contaminated sites in south China. The advantage of intercropping system in the As-contaminated area is that because of the efficient As uptake ability of *P. vittata*, the mobile As in the rhizosphere of the intercropping system decreased, lowering the uptake of As by the intercropping crops (Wan et al. 2017a). Therefore, on the one hand As hyperaccumulator can take up As from soil, and on the other hand, safe products can be produced by the intercropping crops.

Mulberry (*Morus alba* L.) is a deciduous tree species of *Moraceae* widely distributed in Southern China (Chen et al. 2016). Culture of mulberry leaves as feed for silkworm (*Bombyx mori* L.), an economically important insect as the primary producer of silk, has been a traditional planting and processing industry in China and some areas in other Asian countries, contributing to local farming and textile industries (Shi et al. 2016; Singhal et al. 2010). Intercropping of *M. alba* with annual crop is common planting pattern in Asia, to better use the land and solar energy.

Different ecotypes of *P. vittata* showed varied adaptability to intercropping system. Two ecotypes of *P. vittata* were found to be more appropriate for intercropping with *Morus alba* (Fig. 1.11). These two populations showed extensive root overlap with *M. alba* and efficient uptake of bioavailable As, thus depleting As in the rhizosphere and lowering As risk (Wan and Lei 2018). These studies on different *P. vittata* ecotypes have further expanded the application range of this fern (Fig. 1.12).

The As hyperaccumulating ability has been found and confirmed through both lab survey and greenhouse experiment. The biological characteristics indicated this fern species as an excellent plant material for the As contaminated soil phytoextraction technology. Further, some ecotypes were found to have specialties, which can be used to phytoextract multi-metal(loid)s or to intercropping mode.

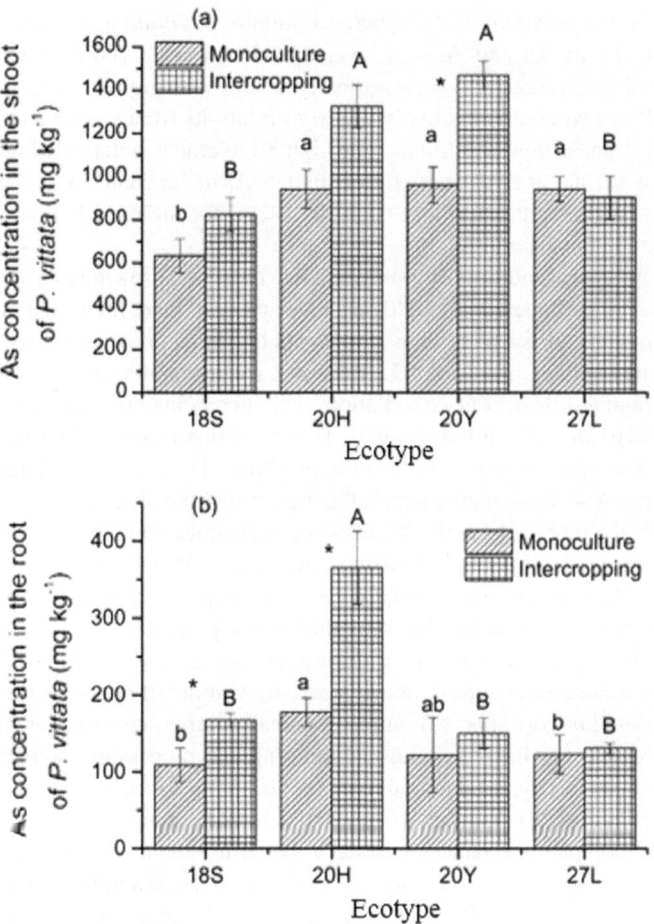

Fig. 1.12 Screening out the *P. vittata* ecotype for intercropping mode (Wan and Lei 2018)

References

An ZZ, Huang ZC, Lei M, Liao XY, Zheng YM, Chen TB (2006) Zinc tolerance and accumulation in *Pteris vittata* L. and its potential for phytoremediation of Zn- and As-contaminated soil. Chemosphere 62:796–802

Chen HW, He XH, Liu Y, Li J, He QY, Zhang CY, Wei BJ, Zhang Y, Wang J (2016) Extraction, purification and anti-fatigue activity of gamma-aminobutyric acid from mulberry (*Morus alba* L.) leaves. Sci Rep 6:10

Chen TB, Wei CY (2000) Arsenic hyperaccumulation in some plant species in South China. In: Luo YM, Cao ZH, Chen YX, McGrath SP, Zhao FJ, Xu JM (eds) SoilRem 2000: international conference of soil remediation. Institute of Soil Science, Chinese Academy of Sciences, Hangzhou, China, pp 194–195

Chen TB, Wei CY, Huang ZC, Huang QF, Lu QG, Fan ZL (2002) Arsenic hyperaccumulator *Pteris Vittata* L. and its arsenic accumulation. Chin Sci Bull 47:902–905

Chen TB, Zhang BC, Huang ZC, Liu YR, Zheng YM, Lei M, Liao XY, Piao SJ (2005) Geographical distribution and characteristics of habitat of As-hyperaccumulator *Pteris vittata* L. in China (in Chinese, abstract in English). Gograph Res 24:825–833

Cunningham SD, Berti DR (1993) Phytoremediation of contaminated soils—progress and promise. Abstracts of Papers of the American Chemical Society 205, 80-PETR

Feng RW, Liao GJ, Guo JK, Wang RG, Xu YM, Ding YZ, Mo LY, Fan ZL, Li NY (2016) Responses of root growth and antioxidative systems of paddy rice exposed to antimony and selenium. Environ Exp Bot 122:29–38

Huang Z, Zhao F, Hua JF, Ma ZH (2018) Prediction of the distribution of arbuscular mycorrhizal fungi in the metal(loid)-contaminated soils by the arsenic concentration in the fronds of *Pteris vittata* L. J Soils Sediments 18:2544–2551

Kramer U (2010) Metal hyperaccumulation in plants. Annu Rev Plant Biol 61:517–534

Ma LQ, Komar KM, Tu C, Zhang WH, Cai Y, Kennelley ED (2001) A fern that hyperaccumulates arsenic—a hardy, versatile, fast-growing plant helps to remove arsenic from contaminated soils. Nature 409:579

Mullner K, Daus B, Mattusch J, Vetterlein D, Merbach I, Wennrich R (2013) Impact of arsenic on uptake and bio-accumulation of antimony by arsenic hyperaccumulator *Pteris vittata*. Environ Pollut 174:128–133

Salt DE, Blaylock M, Kumar N, Dushenkov V, Ensley BD, Chet I, Raskin I (1995) Phytoreme-diation—a novel strategy for the removal of toxic metals from the environment using plants. Bio-Technology 13:468–474

Salt DE, Smith RD, Raskin I (1998) Phytoremediation. Annu Rev Plant Physiol Plant Mol Biol 49:643–668

Shi SM, Chen K, Gao Y, Liu B, Yang XH, Huang XZ, Liu GX, Zhu LQ, He XH (2016) Arbuscular mycorrhizal fungus species dependency governs better plant physiological characteristics and leaf quality of Mulberry (*Morus alba* L.) Seedlings. Front Microbiol 7:11

Singhal BK, Baqual MF, Khan MA, Bindroo BB, Dhar A (2010) Leaf surface scanning electron microscopy of 16 mulberry genotypes (Morus SPP.) with respect to their feeding value in silkworm (*Bombyx mori* L.)rearing. Chil J Agric Res 70:191–198

Tisarum R, Lessl JT, Dong XL, de Oliveira LM, Rathinasabapathi B, Ma LQ (2014) Antimony uptake, efflux and speciation in arsenic hyperaccumulator *Pteris vittata*. Environ Pollut 186:110–114

Wan X, Lei M, Chen T, Yang J (2017a) Intercropped *Pteris vittata* L. and *Morus alba* L. presents a safe utilization mode for arsenic-contaminated soil. Sci Total Environ 579:1467–1475

Wan XM, Lei M (2018) Intercropping efficiency of four arsenic hyperaccumulator *Pteris vittata* populations as intercrops with *Morus alba*. Environ Sci Pollut Res 25:12600–12611

Wan XM, Lei M, Chen TB (2016) Interaction of As and Sb in the hyperaccumulator *Pteris vittata* L.: changes in As and Sb speciation by XANES. Environ Sci Pollut Res 23:19173–19181

Wan XM, Lei M, Chen TB, Zhou GD, Yang J, Zhou XY, Zhang X, Xu RX (2014) Phytoremediation potential of *Pteris vittata* L. under the combined contamination of As and Pb: beneficial interaction between As and Pb. Environ Sci Pollut Res 21:325–336

Wan XM, Lei M, Huang ZC, Chen TB, Liu YR (2010) Sexual propagation of *Pteris vittata* L. Influenced by pH, Calcium, and temperature. Int J Phytorem 12:85–95

Wan XM, Lei M, Liu YR, Huang ZC, Chen TB, Gao D (2013) A comparison of arsenic accumulation and tolerance among four populations of *Pteris vittata* from habitats with a gradient of arsenic concentration. Sci Total Environ 442:143–151

Wan XM, Lei M, Yang JX (2017b) Two potential multi-metal hyperaccumulators found in four mining sites in Hunan Province, China. CATENA 148:67–73

Wu FY, Deng D, Wu SC, Lin XG, Wong MH (2011) Arsenic tolerance, uptake, and accumulation by nonmetallicolous and metallicolous populations of *Pteris vittata* L. Environ Sci Pollut Res 22:8911–8918

Wu FY, Leung HM, Wu SC, Ye ZH, Wong MH (2009) Variation in arsenic, lead and zinc tolerance and accumulation in six populations of *Pteris vittata* L. from China. Environ Pollut 157:2394–2404

Wu FY, Zhakypbek Y, Bi YL, Chen ST, Guo YF, Wong MH (2013) Effects of Lead and Zinc on Arsenic accumulation in nonmetallicolous and metallicolous populations of *Pteris vittata* L. Commun Soil Sci Plant Anal 44:2839–2851

Xiao XY, Chen TB, An ZZ, Lei M, Huang ZC, Liao XY, Liu YR (2008) Potential of *Pteris vittata* L. for phytoremediation of sites co-contaminated with cadmium and arsenic: the tolerance and accumulation. J Environ Sci 20:62–67

Zhang FS, Li L (2003) Using competitive and facilitative interactions in intercropping systems enhances crop productivity and nutrient-use efficiency. Plant Soil 248:305–312

Chapter 2
Arsenic Hyperaccumulation Mechanisms: Absorption, Transportation and Detoxification

Abstract The mechanism for arsenic (As) uptake by the hyperaccumulator *Pteris vittata* and its ability to tolerate high As concentrations is an active research field. The potential As hyperaccumulation mechanisms of *P. vittata* was summarized from the aspects of the distribution and transformation of As in *P. vittata*, the relationship between As and phosphorus (P), the driving force for the As translocation upwards, and the compartmentalization and chelation of As in *P. vittata*. It has been identified that the synergic absorption of As and P, the reduction of As in the endodermis of rhizoid, transportation-driven translocation upwards, and the compartmentation of As in trichrome, and the chelation of As with sulfur have contributed to the efficient hyperaccumulation of As by *P. vittata*. Besides, As is a redox-sensitive element whose toxicity largely depends on its oxidation state and chemical speciation. Synchrotron radiation based X-Ray technologies provide an advanced tool to accurately reflect the existence form and the transferring process.

Keywords Chelation · Compartmentation · Phosphorus · Redox · X-ray fluorescence spectroscopy · X-ray absorption fine structure

2.1 Distribution and Transformation of as in *Pteris vittata*

Since *Pteris vittata* L. was identified as an As hyperaccumulator, it has become a model plant for the study of hyperaccumulating mechanism. The concentration of As in *P. vittata* is tens of thousands of times higher than that of common plants, with the highest concentration of As in the leaves (fronds) reaching 10,000 mg kg^{-1}. Studying the distribution and transformation of As in *P. vittata* can help understand its hyperaccumulation and tolerance mechanisms. The root absorption, translocation from root to shoot, the storage and detoxification of As in *P. vittata* are regarded as processes essential for the As hyperaccumulation by *P. vittata* (Kertulis et al. 2005; Tu and Ma 2005; Wang et al. 2011).

The distribution of As in *P. vittata* follows the order of plume > petioles > roots, and 75–80% of arsenic was stored in the leaves (Chen et al. 2002b). The As concentration in the new leaves of *P. vittata* was the highest when the concentration of arsenic added

© The Author(s) 2020

T. Chen et al., *Phytoremediation of Arsenic Contaminated Sites in China*,
SpringerBriefs in Environmental Science,
https://doi.org/10.1007/978-981-15-7820-5_2

was less than 30 mg kg^{-1}, followed by the mature leaves and the old leaves, while the As concentration of old leaved increased when the concentration of arsenic was more than 200 mg kg^{-1}. Under As stress, more As was accumulated in the senescing fronds (47%) and mature fronds (11%) than the young fronds. In senescing fronds, As concentrations in pinna margin were 2.3 times of the midribs, consistent with As-induced necrotic symptom (Han et al. 2020).

Studying the elements micro-distribution in As hyperaccumulator further reveals the uptake and translocation characteristics of As and other elements at a cellular level, which help understand the arsenic accumulation and tolerance mechanisms. Synchrotron radiation X-ray fluorescence spectroscopy (SR-XRF) had higher detective sensitivity, so it could be used to analyze the distribution of arsenic, which was usually at the level of mg kg^{-1} or even less in plants. SR-XRF scanning of the micro-distribution of As in *P. vittata* was performed at an XRF station at beamline 4W1B of the Beijing Synchrotron Radiation Facility. The electron storage ring was operated at 2.2 GeV with the current ranging from 59 to 114 mA. Samples were fixed in a high precision sample positioning stage, with 1 μm per step in three dimensions, driven by computer-controlled stepping motors. The sample profile was at a 45° angle with the beam direction and the distance between the sample and detector fixed at 50 mm. Fluorescent radiation was detected using a PGT Si (Li) solid detector with a 7.5 μm-thick beryllium window. Spectra were analyzed using WinQXAS software and peak areas calculated after background deduction. Fluorescence intensity was normalized based on the electron current and used for quantitative comparison.

SR-XRF results indicated that the concentration of As in the veins was higher than that on both sides, and the biggest difference appeared at the tip of the leaves (Chen et al. 2005a). It implied that *P. vittata* had the ability to transfer As from xylem to mesophyll. Through energy dispersive X-ray microanalyses (EDXA), it has been found that As mainly distributed in the upper epithelial cells of the feathers (Lombi et al. 2002). Using SRXRF, Chen et al. (2003) further indicated that on the cellular level, As was mainly distributed in mesophyll tissue (fence tissue), while concentration in epidermal cells was relatively low. At the sub-cellular level, through centrifugal separation it has been found that a large amount of As accumulated in the cytosol of the feathers, which reached 78% of the As in the leaves (Fig. 2.1), and the As storage capacity of the cell wall was only 17% (Chen et al. 2005b).

The distributions of As and six essential elements in the pinna of *P. vittata* were studied using SRXRF. The distribution of As in the pinna showed that As had great abilities to be transported in xylem vessels and from xylem to mesophyll (Chen et al. 2005a). The distribution of As and the other 6 elements is shown in Fig. 2.2 with the darker color denoting the higher concentrations of the elements at the point scanned. Different distribution pattern among elements could be found in the maps. The distribution of K, one of the most mobile elements in plants, was similar to that of As. The distributions of manganese (Mn) and zinc (Zn) in the direction parallel to midrib are similar to that of As while their ability of transportation from xylem to mesophyll was poorer than that of As and potassium (K). The distributions of less mobile iron (Fe) and calcium (Ca) are almost opposite to that of As in pinna and they are mainly located in the region close to base.

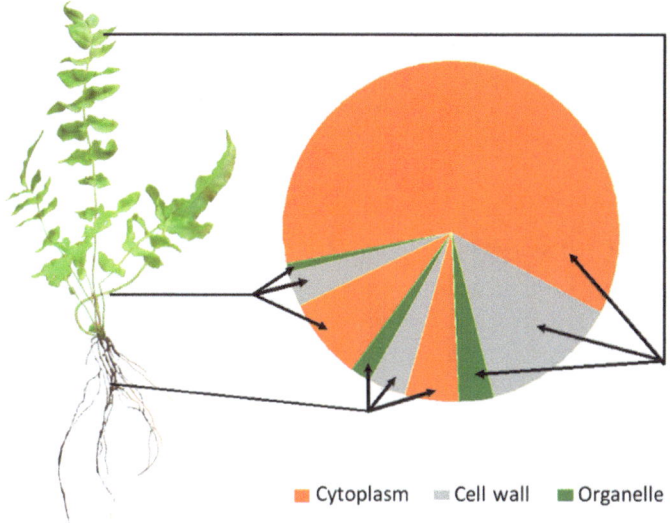

Fig. 2.1 Subcellular distribution of As in *P. vittata* (Chen et al. 2005b)

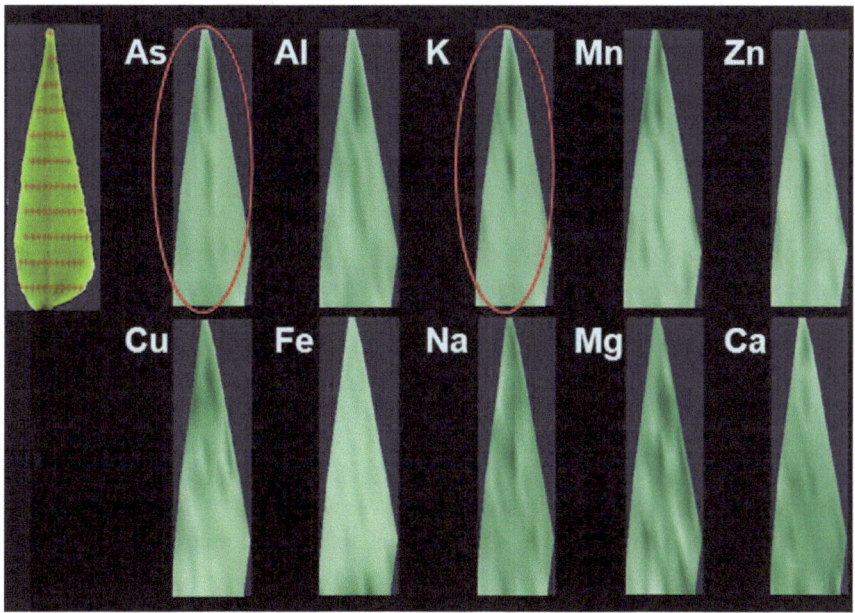

Fig. 2.2 Distribution of As and other elements on the plume of *P. vittata*

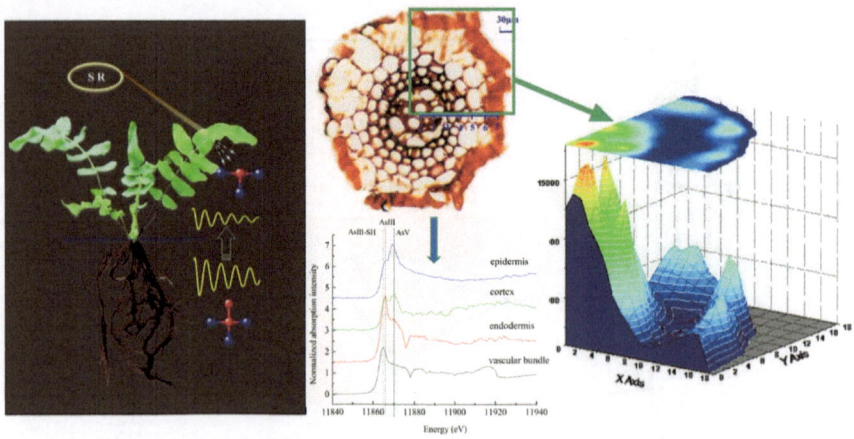

Fig. 2.3   Disclosure of the distribution, speciation and migration of arsenic in hyperaccumulator

The species of As had great effects on its uptake and translocation in *P. vittata*. The study on the transformation of As in *P. vittata* and its impact on the As uptake by the hyperaccumulator helped understanding how hyperaccumulators accumulate As and defend themselves against the arsenic toxicity. Most methods used for As speciation measurement, such as High Performance Liquid Chromatography coupled Inductively Coupled Plasma—Mass Spectrometry (HPLC-ICP-MS), require complicated pretreatment procedures, such as extraction and separations, which may alter the As species.

Synchrotron radiation based extended X-ray absorption fine structure (SR-EXAFS) provides a method to directly determine the elemental oxidation state and local coordination environments in the samples without any chemical pretreatments (Fig. 2.3). SR-EXAFS has been confirmed as a promising tool to acquire the actual information about elemental species and transformation in plant, and is growing up to be an important technique for researches in the environmental and biological field.

Using SR-EXAFS, the arsenic species and transformation during root uptake and translocation processes in *P. vittata* have been studied (Fig. 2.4), to reveal the mechanisms of arsenic hyperaccumulation and tolerance (Huang et al. 2004). Arsenic mainly exists in As(V) at the root, while As(III) is rare, accounting for only 8.3% of the total As. Some studies found that As(V) was quickly converted to As(III) after being absorbed by the root of *P. vittata*, and this transformation process mainly occurred in the root (Huang et al. 2004).

The study on As species and transformation in *P. vittata* by EXAFS technique shows that some of the As in the root of *P. vittata* coordinated with sulfur, but most of As in the pinna was coordinated with oxygen. As(V) was reduced to As(III) after it was absorbed, and kept as As(III)-O in the above-ground tissues of *P. vittata* (Fig. 2.4). The coordination of As and sulfur (S) in *P. vittata* was similar to that from As-GSH, meaning that some As in *P. vittata* existed as As-GSH. But the results from HPLC-ICP-MS (Fig. 2.3) showed that the arsenic in *P. vittata* mainly presented as inorganic

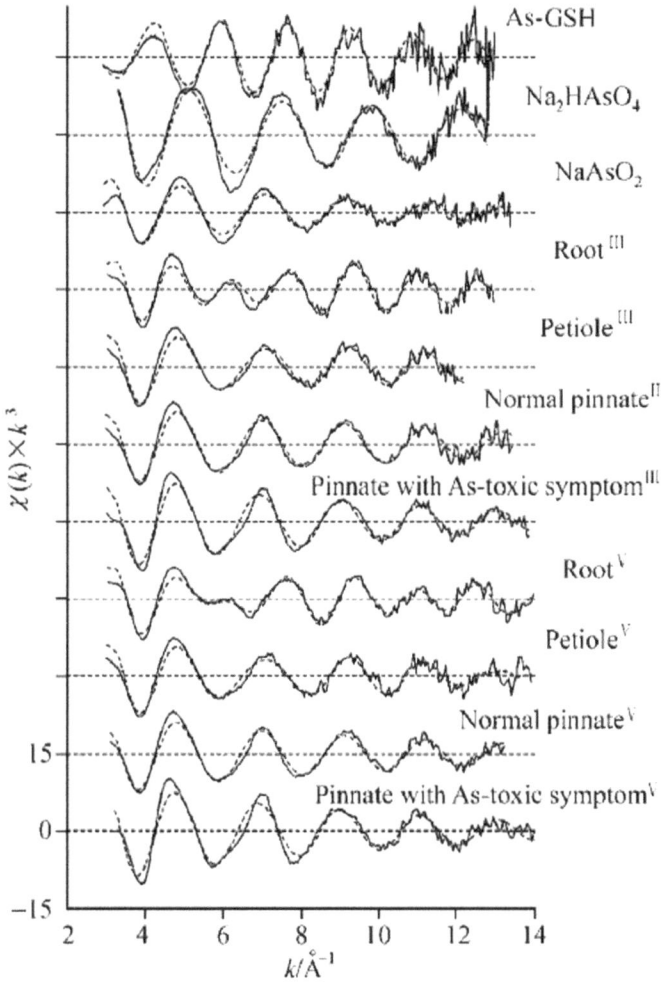

Fig. 2.4 Speciation of As in *P. vittata* disclosed by SR-EXAFS (Huang et al. 2004)

arsenic. This difference may result from the complicated pretreatment procedures required for HPLC-ICP-MS, like grinding to break plant cells and extraction with chemical reagent.

With the development of SR based X-Ray absorption and fluorescence technologies, the spatial and time dissolution of this method was greatly improved. SR based micro-XANES was not only used to determine the As species in bulk samples, but also used to disclose the species of As at the micro level in different tissues of the rhizoid (Fig. 2.5), petiole, and pinna of *P. vittata* (Wan et al. 2017). Scanning were performed with a $3.18 \times 2.56 \ \mu m^2$ spatial resolution using a Kirkpatrick-Baez

Fig. 2.5 Example spectra of micro-XANES spectra of *P. vittata* tissue: As(III) exposed root. The solid line is the experimental data, the dotted line is the fit of the data (Wan et al. 2017)

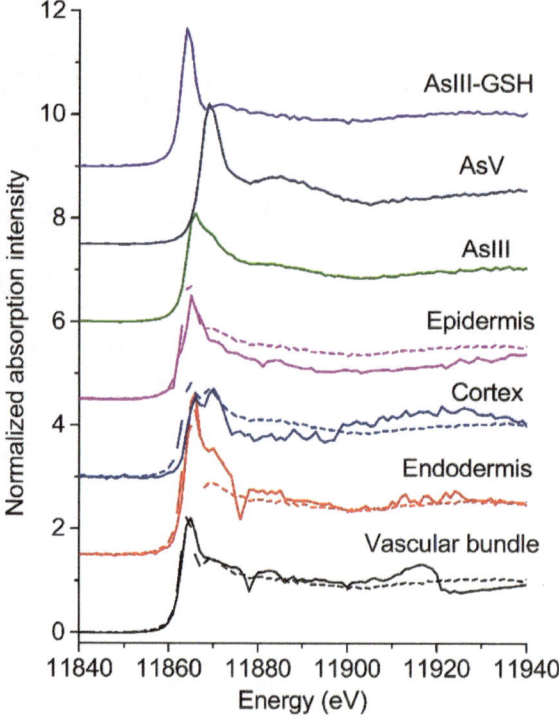

focusing mirror. The step size of energy was 1 eV from 11,840 to 11,900 eV and 2 eV after 11,900 eV. The data was normalized and analyzed using the IFEFFIT program.

The reduction of AsV into AsIII, oxidation of AsIII into AsV, and chelation of AsIII with thiols were all observed in *P. vittata*. The reduction of As mainly occurred in the rhizoid, whereas oxidation and chelation mainly occurred in the aboveground parts. Correlation analysis showed that the As concentration in pinna was significantly correlated with the AsV percentage in paraxial and abaxial epidermis (positive), AsIII-GSH percentage in paraxial epidermis (positive), and AsIII percentage in paraxial and abaxial epidermis (negative). Both oxidation and chelation reactions contributed to the accumulation of As in *P. vittata*. In general, micro-XANES is a powerful method to determine arsenic oxidation states of plant tissues at the cellular level, enabling the speciation analysis with high spatial resolution and low detection limits (Bacquart et al. 2007, 2010), which sometimes can reflect the As species that is hard to recognize using bulk-XANES.

Recently, 'depth information' (3D) of As chemical speciation in different cell types within tissues of *P. vittata* has been disclosed using confocal volumetric XAS. Differences have been observed in As speciation across tissues. Arsenic distribution is linked to K in ion channels of the phloem bundles (van der Ent et al. 2020).

2.2 Relationship of Arsenic with Phosphorus

The elements As and phosphorus (P) both belong to Group 15 on the Periodic Table of Elements, thus having some similar chemical characteristics. However, these two elements seem to behave differently in plants. P is an essential nutrient for plant growth whereas As is a toxic trace element not necessary for plant growth. Studies have been conducted to investigate the relationship between As and P in the system of soil and As hyperaccumulator *P. vittata*.

Earlier hydroponic experiments disclosed an inhibition effect of P on the uptake of As by *P. vittata*, and suggested that such inhibition might result from the utilization of P transporter by As (Tu and Ma 2003; Wang et al. 2002). However, pot experiments showed that low concentration of P (<400 mg kg^{-1}) had no significant effect on the As concentration of *P. vittata*, and the bioconcentration factor of As. But the addition of a large amount of phosphorus (400 mg kg^{-1} or more) increased the As concentration in the aboveground parts and roots of hyperaccumulator *P. vittata*, and the bioconcentration factor of As (Chen et al. 2002a). It seems that there is no antagonistic effect between As and P. On the contrary, there is synergistic effect between these two elements at a high concentration.

The interaction between As and P was further investigated in a dynamic and continuous way employing a synchrotron X-ray microprobe (Lei et al. 2012). Special attention was paid to the interaction between As and P in different tissues of *P. vittata* rhizoids, especially to the endodermis, where arsenate was reduced to arsenite, disclosed by Micro XANES (Fig. 2.6).

Figure 2.6a indicated that P deficiency increased the content of As in *P. vittata* rhizoids and altered the As distribution pattern along the passway from epidermis to vascular bundle. Compared to the normal P supply, As concentrations in rhizoids with no P supply were higher. And such increase pattern was uneven in different tissues. After 12 h of As exposure ($-$P $+$ As-12 h), As content in the epidermis and exodermis was ~60% higher than that under normal P supply, while the increase in As content in the endodermis and vascular bundle was not apparent.

With the extended As exposure, As in the epidermis, exodermis, and cortex increased while As in the endodermis and vascular bundle showed no similar trends, indicating a possible blockage of As transport in the cortex and endodermis, similar to that under normal P supply.

The P distribution and dynamic changes in *P. vittata* rhizoids were different from that of As. After 0.5 h of As exposure ($-$P $+$ As-0.5 h), P was highest in the vascular bundle, opposite to the result with a normal P supply and, with the extended As exposure, P contents in the 6 scanned sites decreased (Fig. 2.6b). The exposure of *P. vittata* to As exposure may have acted as a trigger for enhancing P transfer from the epidermis to vascular bundle until there was no longer available P from the epidermis (caused by P deficiency), which was different from the observed As detention in the epidermis with As absence.

As and P displayed great similarity during the transportation process from epidermis to cortex of *P. vittata*, with the addition or absence of As or P, implied

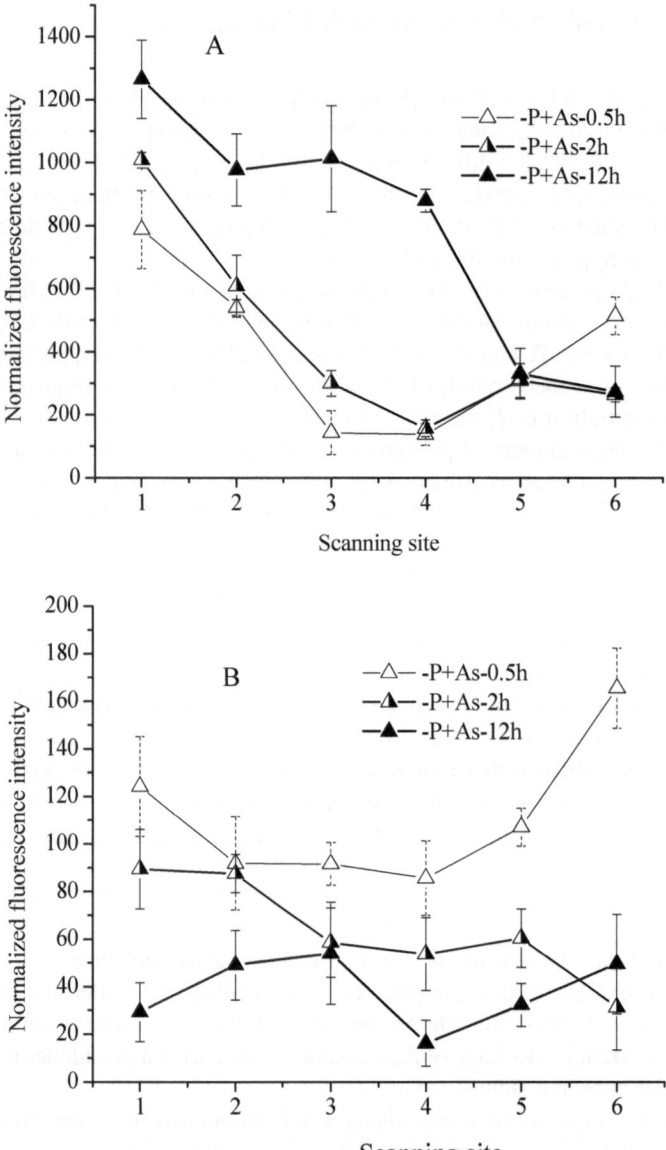

Fig. 2.6 Dynamic uptake of As and P under P deficiency (A, As uptake; B, P uptake; 1, epidermis; 2, exodermis; 3–4, cortex; 5, endodermis; 6, vascular bundle) (Lei et al. 2012)

the co-transport of arsenate and phosphate in *P. vittata* rhizoids. However, during the transportation from endodermis to vascular bundle, arsenite were no longer co-transported with P, detected from the influence of P deficiency and energy inhibition on the transport to two elements. The efficient transport of arsenite in As hyperaccumulator should be paid more attention.

2.3   The Translocation of as Driven by Transpiration

It has been found that the cytoplasmic supernatant of *P. vittata*'s pinnae is the major place storing As taken up from growth media, reaching up to 60% of total As in the whole seedling (Chen et al. 2005b). The particularly efficient translocation is one of the most important processes during the uptake of As by *P. vittata* (Poynton et al. 2004; Su et al. 2008). Mineral ions enter into the xylem of plant roots by apoplastic method or symplastic method, and then are generally transported upwards driven by transpiration (Hinsinger 1998; Russell and Shorrocks 1959). The method used by the hyperaccumulators to transport trace elements is an active research topic.

Different from normal plants that often reduce water uptake when confronting heavy metal stress (Kholodova et al. 2011; Vernay et al. 2007), As hyperaccumulator increased water uptake under As stress (Lee and Lee 2011). Therefore, it was suggested that the water metabolism may affect the As uptake by hyperaccumulators (Doucleff and Terry 2002; Kabata-Pendias and Pendias 2001). A recent study found that the translocation of As in *P. vittata* involved 3 steps: (1) the movement of As(V) (prevailing species in the rhizosphere) using P pathway from epidermis to endodermis; (2) the reduction of As(V) to As(III) in endodermis; and (3) the transport of As(III) afterwards via a more efficient process, likely related to water metabolism (Lei et al. 2012).

P. vittata has huge biomass (~36 t dry weight hm^{-2} y^{-1}) and very developed aboveground parts. It is suggested that transpiration might be the main upwards translocation driving force of As after the reduction of As(V) to As(III). Aiming to testify this hypothesis, the As uptake pattern in *P. vittata*'s fronds under different As exposure concentrations and transpiration altering treatments was disclosed (Wan et al. 2015).

Therefore, transpiration made a notable contribution to As uptake by *P. vittata*, based on the observation that: (a) with an increase in the exposure concentration of As, the shoot As concentration of *P. vittata* increased proportionally; (b) inhibition of transpiration decreased As concentrations in shoots while improving transpiration increased As concentrations in shoots; (c) *P. vittata* ecotype with higher transpiration rate accumulated more As in the aboveground parts than that with lower transpiration rate.

Comparison between the measured As amount and calculated As amount (multiplying the water loss and the As concentration in the culture solution) was conducted. Calculated As accounted for more than half of the measured As amount (Fig. 2.7).

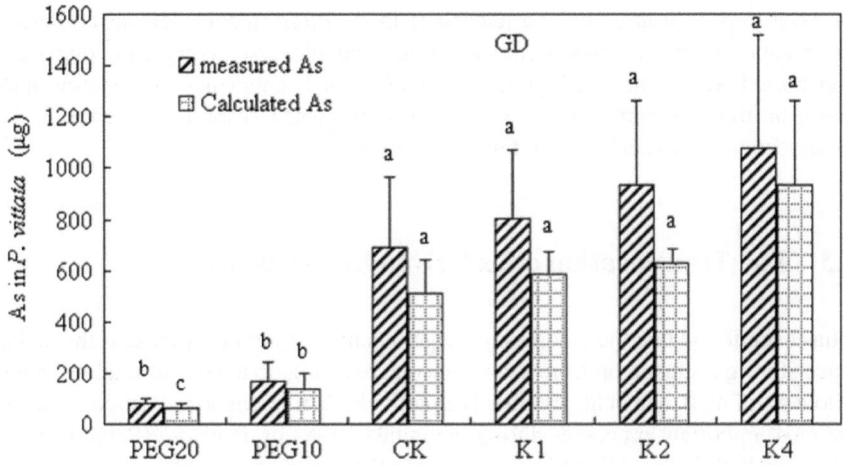

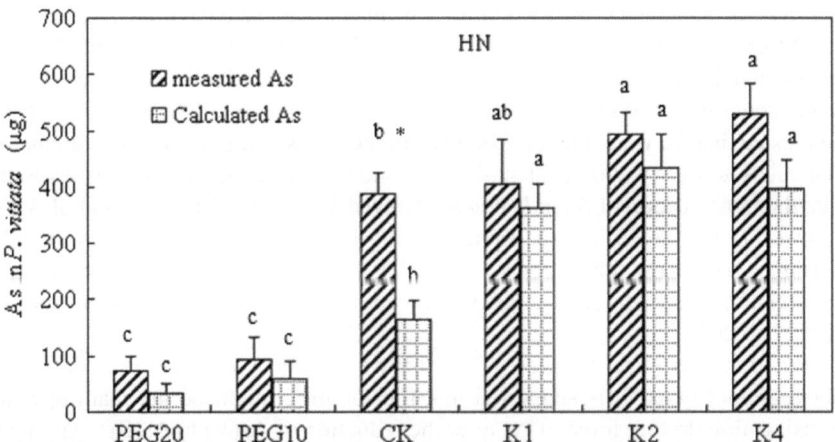

Fig. 2.7 Measured and Calculated As amount. Measured As amount = Shoot As concentration of *P. vittata* × shoot biomass. Calculated As amount = As concentration of the culture solution × water loss. GD and HN indicates Guangdong and Hunan ecotype of *P. vittata*. CK is the control with *P. vittata* growing in the normal culture solution. PEG 10 and PEG 20 indicate the transpiration decreasing treatments, created by adding 10 and 20% Polyethylene glycol 6000 to the culture solution, respectively. K1-K4 indicate the transpiration enhancement treatments, created by adding 1–4 mM KNO_3 to the culture solution. Different letters beside error bar indicate significant difference among treatments. Asterisks indicate significant differences between measured As and calculated As in *P. vittata* ($P < 0.05$, $n = 4$). (Wan et al. 2015)

Transpiration accounted for ~74 and ~50% of the As accumulated by GD and HN ecotype, respectively.

Lei et al. suggested that before entering into the endodermis of *P. vittata* rhizoids, As utilizes phosphate pathway while after that, phosphate pathway no longer works (2012), which means that the transport of As after endodermis requires another carrier. It is further proposed that this carrier is likely to be aquaglyceroporins, which is essential for the water uptake in plants (Luu and Maurel 2005; Preston et al. 1992). However, until now there has no direct evidence confirming the role of aquaglyceroporins in the uptake of AsIII.

The suppression of transpiration obviously decreased As concentration at the edge of *P. vittata* pinnae, implying the role of transpiration in As uptake and distribution in *P. vittata*. Inhibiting transpiration blocked As's move from vein to epidermis, indicating that the lack of enough transpiration as a drive force disabled the transfer of As from vein to epidermis, which is one of the main places to store As in *P. vittata*. And there existed obvious similarity in the distribution in pinnae between K and As, implying that two elements are high likely transported together. In contrast, the distribution of P had no similarity with As, indicating that two elements may no longer share carriers in pinnae.

As uptake and transpiration displayed strong correlation. Higher transpiration due to transpiration regulation or ecotypic difference always led to higher As uptake by *P. vittata*. And on the other hand, low As exposure greatly increased the transpiration of *P. vittata*, similar to another As hyperaccumulator *P. cretica* (Lee and Lee 2011). Transpiration was found to be the main drive for *P. vittata* to accumulate and re-distribute As in pinnae.

2.4 Compartmentation and Chelation with Thiols

Other than the As absorption, As transformation, and As translocation, there is another mechanism that is also a hot research topic: the As detoxification.

Environmental scanning electron microscope (ESEM) fitted with an energy dispersive X-ray micro analyzer (EDX) was used to investigate the surface micro-morphology and As micro-distribution in *P. vittata*. It was found that amounts of trichome, which possessed multicellular structure with the average length of 160 μm and with an average diameter of 28 μm, existed in the frond of *P. vittata*, and the density of trichome on the pinnate axial surface was higher than that on the petiole.

Visible X-ray peak of As was recorded in the epidermal cell and trichome. The relative weight of As in the pinnate trichome, which contained the highest concentration of As among all tissues of the plant (Fig. 2.8), was 2.4 and 3.9 times as much as that in the cells of epidermis and mesophyll, respectively (Li et al. 2005).

Furthermore, the As concentrations in the basal and stalk cells of the same trichome were higher than that in its cap cell. Results indicated that the trichome of *P. vittata* plays an important role in As hyperaccumulation.

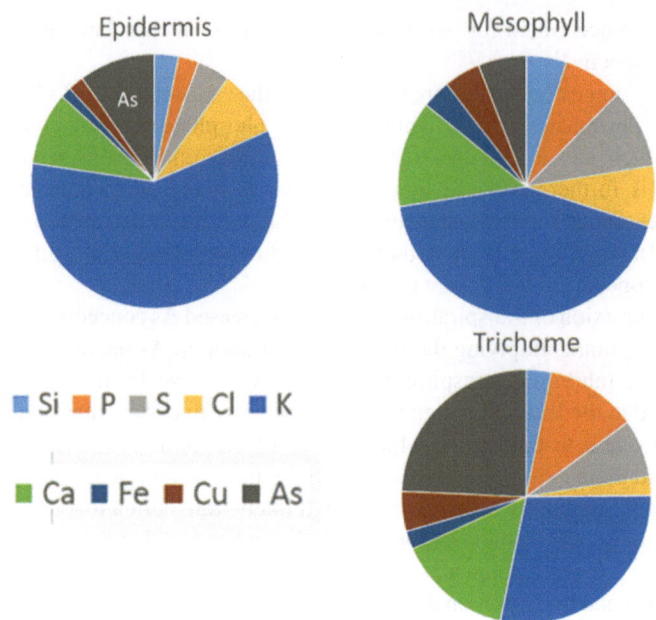

Fig. 2.8 The contents of elements in different tissues of *P. vittata*

Different from the distribution on pinnae, K no longer showed similarity with As, in terms of the distribution among tissues. In stead, P showed similarity with As. There was a high percentage of P in trichome, apparently higher than that of epidermis and mesophyll.

Cytoplasmic supernatant fraction is another place for the As compartmentation in *P. vittata* (Chen et al. 2005b). When the plant was grown in a nutrient solution without additional As, most of the accumulated As was isolated to the cell wall. However, in plants growing in a nutrient solution containing 0.1 or 0.2 mM As, approximately 78% of the total As accumulated within the pinna. The proportions of As accumulation in the cytoplasmic supernatant fraction were 78% of that in the pinna and 61% of that in the plant. In either treatment group (0.1 or 0.2 mM As), the fraction containing the lowest level of As was the organelle fraction. These results suggest that As compartmentalization in the cytoplasmic supernatant fraction may play a role in the detoxification of As in *P. vittata*.

Besides As compartmentalization, comparison between *P. vittata* and non-hyperaccumulator fern indicated that the chelation of As with sulfur (S) is another As detoxification method of *P. vittata*. Phytochelatins (PCs) can chelate with As in *P. vittata*, lowering the concentration of As in form of free ion, therefore reducing the harm of As to plant tissue. With the exposure As concentration increasing from 150 to 300 mg kg^{-1}, thiols and glutathione in *P. vittata* significantly increased (Srivastava et al. 2009). This indicated that PCs synthesis was induced upon the improved exposure of *P. vittata* to AsV. The As concentration of *P. vittata* correlated significantly with PC2 concentration in both roots and shoots (Zhao et al. 2003).

Under As exposure, sulfhydryl groups (-SH) increased in both the As hyperaccumulator *P. vittata* and the As hypertolerant plant *Adiantum capillus-veneris*, indicating that arsenate enhanced sulfur assimilation (Li et al. 2009). In *A. capillus-veneris*, As was mainly coordinated with S; whereas in *P. vittata*, it was coordinated with oxygen (Fig. 2.9). Differences in As concentration and in the rate of As reduction were noted between the two plant species. In *A. capillus-veneris*, As was present at lower levels and was reduced and coordinated with -SH. This was considered to represent a defense strategy to limit As transport to the frond. For *P. vittata*, the SH group might be an electron donor to reduce As (V) to As (III), therefore fewer SH groups were used to coordinate with reduced As (III) (Fig. 2.10). This was regarded as an accumulation strategy to facilitate As transport. The results suggested that S played important roles in As detoxification and accumulation in tolerant and hyperaccumulating plants, respectively.

Further study was conducted on the impacts of the in vivo regulation of sulfur (S) metabolism on the speciation of uptake of arsenic (As) and S by *P. vittata* using SR-XAFS (Lei et al. 2013). The S assimilation inhibitor L-buthionine-sulphoximine (BSO) markedly inhibited As reduction, doubling the concentration of As(V) in *P. vittata* rhizoids. This inhibition of As(V) reduction resulted in an As blockage in the rhizoids, decreasing As transport to aboveground parts by 47%. The significant impact of BSO demonstrated -SH's vital role as a reductant in As(V) reduction

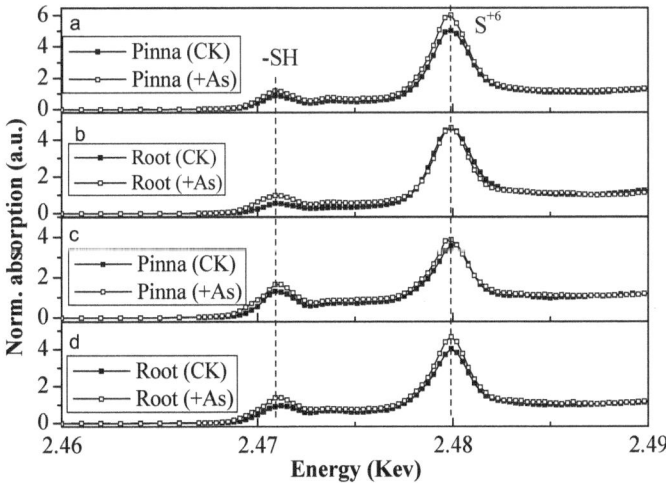

Fig. 2.9 The sulfur K near-edge spectra of Adiantum capillus-veneris (a, b) and *Pteris vittata* (c, d) in the range of 2.46–2.52 keV. The CK is the nutrition solution treatment (no As) and the +As treatment has had 2 mg kg^{-1} Na$_2$HAsO$_4$ added to the nutrition solution (Li et al. 2009)

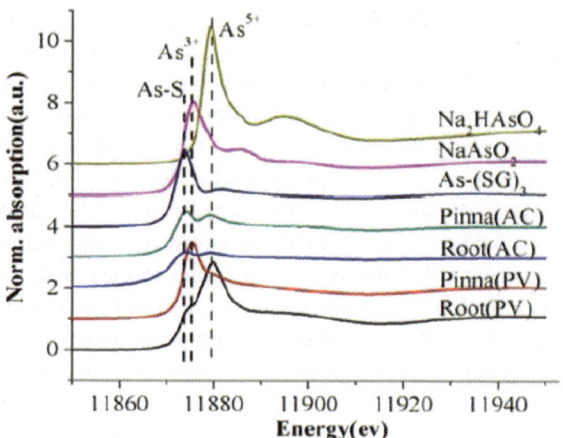

Fig. 2.10 Arsenic XANES spectra from 11,850 to 11,950 eV of the arsenic model compounds and plant samples of *Adiantum capillus-veneris* (AC) and *Pteris vittata* (PV) (Li et al. 2009)

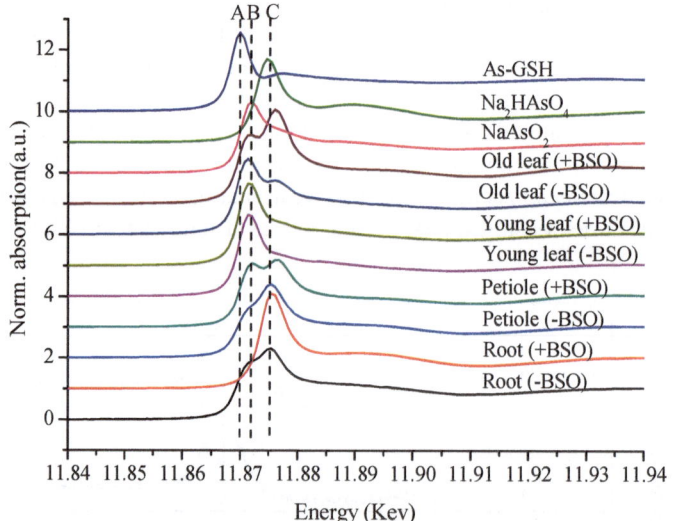

Fig. 2.11 In vivo negative regulation of S metabolism on As species in *P. vittata*: A: GSH-As; B: NaAsO$_2$; C: Na$_2$HAsO$_4$ (Lei et al. 2013)

and confirmed the importance of As(V) reduction in As accumulation in *P. vittata* (Fig. 2.11).

The application of S metabolism accelerant O-acetyl-L-serine (OAS) led to the appearance of large amounts of As-SH in the rhizoids (Fig. 2.12). OAS had no obvious impact on As accumulation but effectively increased the *P. vittata* biomass under As stress, possibly through chelation of excess As with -SH. Thus, -SH may act as both a reductant and a chelator of As in *P. vittata* and the ratio of -SH to As may have determined the specific role of -SH in *P. vittata*.

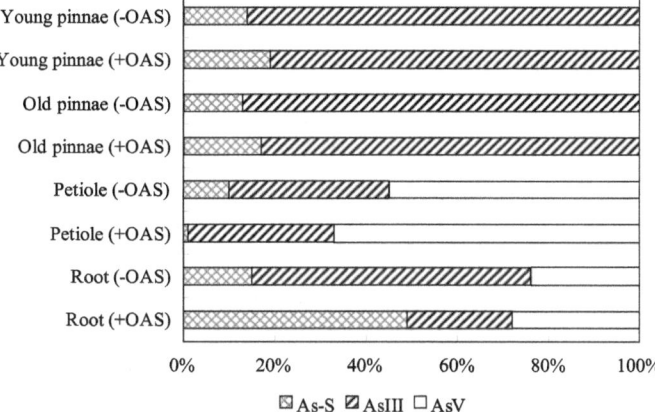

Fig. 2.12 XANES fitting of As species in *P. vittata* with positive S regulation (Lei et al. 2013)

P. vittata is one of the most extensively-studied hyperaccumulator globally to date. As a consequence, much is now known about the uptake and metabolic regulation of As in this species. Nevertheless, many open questions remain. The emerging new advanced technologies may enable the in vivo 3D investigation of the uptake process of As, which can reflect the whole picture more accurately.

References

Bacquart T, Deves G, Carmona A, Tucoulou R, Bohic S, Ortega R (2007) Subcellular speciation analysis of trace element oxidation states using synchrotron radiation micro-X-ray absorption near-edge structure. Anal Chem 79:7353–7359

Bacquart T, Deves G, Ortega R (2010) Direct speciation analysis of arsenic in sub-cellular compartments using micro-X-ray absorption spectroscopy. Environ Res 110:413–416

Chen TB, Fan ZL, Lei M, Huang ZC, Wei CY (2002a) Effect of phosphorus on arsenic accumulation in As-hyperaccumulator *Pteris vittata* L. and its implication. Chin Sci Bull 47:1876–1879

Chen TB, Huang ZC, Huang YY, Lei M (2005a) Distributions of arsenic and essential elements in pinna of arsenic hyperaccumulator *Pteris vittata* L. Sci China Ser C-Life Sci 48:18–24

Chen TB, Wei CY, Huang ZC, Huang QF, Lu QG, Fan ZL (2002b) Arsenic hyperaccumulator *Pteris vittata* L. and its arsenic accumulation. Chin Sci Bull 47:902–905

Chen T, Huang Z, Huang Y, Xie H, Liao X (2003) Cellular distribution of arsenic and other elements in hyperaccumulatorPteris nervosa and their relations to arsenic accumulation. Chin Sci Bull 48(15):1586-1591

Chen TB, Yan XL, Liao XY, Xiao XY, Huang ZC, Xie H, Zhai LM (2005b) Subcellular distribution and compartmentalization of arsenic in *Pteris vittata* L. Chin Sci Bull 50:2843–2849

Doucleff M, Terry N (2002) Pumping out the arsenic. Nat Biotechnol 20:1094–1095

HanY-H, Jia M-R, Wang S-S, Deng J-C, Shi X-X, Chen D-L, Chen Y, Ma LQ (2020) Arsenic accumulation and distribution in *Pteris vittata* fronds of different maturity: impacts of soil As concentrations. Sci Total Environ 715

Hinsinger P (1998) How do plant roots acquire mineral nutrients? Chemical processes involved in the rhizosphere. In: Donald LS (ed) Advances in agronomy. Academic Press, pp 225–265

Huang ZC, Chen TB, Lei M, Hu TD, Huang QF (2004) EXAFS study on arsenic species and transformation in arsenic hyperaccumulator. Sci China Ser C-Life Sci 47:124–129

Kabata-Pendias A, Pendias H (2001) Chapter 5 trace elements in plants Trace elements in soils and plants, 3rd. CRC Press, Boca Raton, pp 73–84

Kertulis GM, Ma LQ, MacDonald GE, Chen R, Chen R, Winefordner JD, Cai Y (2005) Arsenic speciation and transport in *Pteris vittata* L. and the effects on phosphorus in the xylem sap. Environ Exp Bot 54:239–247

Kholodova V, Volkov K, Abdeyeva A, Kuznetsov V (2011) Water status in *Mesembryanthemum crystallinum* under heavy metal stress. Environ Exp Bot 71:382–389

Lee SJ, Lee JP (2011) Effect of arsenic absorption on the water-refilling speed of *Pteris cretica*. Microsc Res Tech 74:517–522

Lei M, Wan X-M, Huang Z-C, Chen T-B, Li X-W, Liu Y-R (2012) First evidence on different transportation modes of arsenic and phosphorus in arsenic hyperaccumulator *Pteris vittata*. Environ Pollut 161:1–7

Lei M, Wan X-M, Li X-W, Chen T-B, Liu Y-R, Huang Z-C (2013) Impacts of sulfur regulation in vivo on arsenic accumulation and tolerance of hyperaccumulator *Pteris vittata*. Environ Exp Bot 85:1–6

Li WX, Chen TB, Chen Y, Lei M (2005) Role of trichome of *Pteris vittata* L. in arsenic hyperaccumulation. Sci China Ser C-Life Sci 48:148–154

Li XW, Lei M, Chen TB, Wan XM (2009) Roles of sulfur in the arsenic tolerant plant adiantum capillus-veneris and the hyperaccumulator *Pteris vittata*. J Korean Soc Appl Biol Chem 52:498–502

Lombi E, Zhao FJ, Fuhrmann M, Ma LQ, McGrath SP (2002) Arsenic distribution and speciation in the fronds of the hyperaccumulator *Pteris vittata*. New Phytol 156:195–203

Luu DT, Maurel C (2005) Aquaporins in a challenging environment: molecular gears for adjusting plant water status. Plant Cell Environ 28:85–96

Poynton CY, Huang JWW, Blaylock MJ, Kochian LV, Elless MP (2004) Mechanisms of arsenic hyperaccumulation in Pteris species: root As influx and translocation. Planta 219:1080–1088

Preston GM, Carroll TP, Guggino WB, Agre P (1992) Appearance of water channels in *Xenopus oocytes* expressing red-cells CHIP28 protein. Science 256:385–387

Russell RS, Shorrocks VM (1959) The relationship between transpiration and the absorption of inorganic ions by intact plants. J Exp Bot 10:301–316

Srivastava M, Ma LQ, Rathinasabapathi B, Srivastava P (2009) Effects of selenium on arsenic uptake in arsenic hyperaccumulator *Pteris vittata* L. Biores Technol 100:1115–1121

Su YH, McGrath SP, Zhu YG, Zhao FJ (2008) Highly efficient xylem transport of arsenite in the arsenic hyperaccumulator *Pteris vittata*. New Phytol 180:434–441

Tu C, Ma LQ (2005) Effects of arsenic on concentration and distribution of nutrients in the fronds of the arsenic hyperaccumulator *Pteris vittata* L. Environ Pollut 135:333–340

Tu S, Ma LQ (2003) Interactive effects of pH, arsenic and phosphorus on uptake of As and P and growth of the arsenic hyperaccumulator *Pteris vittata* L. under hydroponic conditions. Environ Exp Bot 50:243–251

van der Ent A, de Jonge MD, Spiers KM, Brueckner D, Montarges-Pelletier E, Echevarria G, Wan X-M, Lei M, Mak R, Lovett JH, Harris HH (2020) Confocal volumetric mu XRF and fluorescence computed mu-tomography reveals arsenic three-dimensional distribution within intact *Pteris vittata* Fronds. Environ Sci Technol 54:745–757

Vernay P, Gauthier-Moussard C, Hitmi A (2007) Interaction of bioaccumulation of heavy metal chromium with water relation, mineral nutrition and photosynthesis in developed leaves of *Lolium perenne* L. Chemosphere 68:1563–1575

Wan X, Lei M, Chen T, Ma J (2017) Micro-distribution of arsenic species in tissues of hyperaccumulator *Pteris vittata* L. Chemosphere 166:389–399

Wan XM, Lei M, Chen TB, Yang JX, Liu HT, Chen Y (2015) Role of transpiration in arsenic accumulation of hyperaccumulator *Pteris vittata* L. Environ Sci Pollut Res 22:16631–16639

Wang JR, Zhao FJ, Meharg AA, Raab A, Feldmann J, McGrath SP (2002) Mechanisms of arsenic hyperaccumulation in *Pteris vittata*. Uptake kinetics, interactions with phosphate, and arsenic speciation. Plant Physiol 130:1552–1561

Wang X, Ma LQ, Rathinasabapathi B, Cai Y, Liu YG, Zeng GM (2011) Mechanisms of efficient arsenite uptake by arsenic hyperaccumulator *Pteris vittata*. Environ Sci Technol 45:9719–9725

Zhao FJ, Wang JR, Barker JHA, Schat H, Bleeker PM, McGrath SP (2003) The role of phytochelatins in arsenic tolerance in the hyperaccumulator *Pteris vittata*. New Phytol 159:403–410

Chapter 3
Establishment of Phytoremediation Technology for Arsenic Contaminated Soil

Abstract Arsenic (As) contaminated soil is a major issue in China. The discovery of As hyperaccumulator *Pteris vittata* L. opens a door for the phytoextraction of arsenic contaminated soils. The phytoremediation technology includes selecting proper *P. vittata* ecotype, reproduction and transplantation of *P. vittata* sporelings, fertilizer and chemical amendments, harvest and disposal. Depending on the local environmental conditions, scenarios for the irrigation, harvest time and frequency, and the disposal of As-enriched biomass were established. During field practices, the germination and cultivation of *P. vittata* sporelings, and the safe disposal of As–enriched *P. vittata* biomass were identified as essential processes. Therefore, fast reproduction technology of *P. vittata* sporeling, and safe incineration technology were paid focused attention. Furthermore, strengthening measures for the As phytoremediation technologies were proposed from the aspects of fertilizer application, harvest frequency, and ecotype selection. Finally, the biogeochemical cycles of As in the phytoremediation system was compared with the system with normal plants growing. It indicated that the application of phytoremediation led to an As net output of 7.02 kg ha^{-1} A^{-1}, while on the field with normal plant growing, As was accumulating in soil at a rate of 0.24 kg ha^{-1} A^{-1}.

Keywords Arsenic · Biogeochemical cycles · Field practice · Phytoremediation

3.1 Frames of the Phytoremediation Technology of Arsenic Contaminated Soil

Arsenic (As) is a toxic metalloid, and it ubiquitously exists in environment as a result of natural and human activities. Chronic exposure to As through drinking water and staple foods is a major health concern worldwide (Eisler 1985; Sturchio et al. 2013).

Considering the wide distribution and high toxicity of As, many efforts have been devoted to investigate suitable remediation schemes for As contaminated soil. Phytoextraction has increasingly attracted attention, since it is environmental friendly, of simple operation and economic efficiency (Chen et al. 2002; Sirkin et al. 2011). In situ phytoextraction projects using As-hyperaccumulator *P. vittata* L. have

© The Author(s) 2020   33
T. Chen et al., *Phytoremediation of Arsenic Contaminated Sites in China*,
SpringerBriefs in Environmental Science,
https://doi.org/10.1007/978-981-15-7820-5_3

been established on farmlands and residential areas in China and America, with a high As removal rate achieved (Ebbs et al. 2010; Huang et al. 2007; Wan et al. 2016).

The general procedure of a phytoremediation project using the hyperaccumulator *P. vittata* is shown in Fig. 3.1. Firstly, the pollution status of the soil was assessed, based on which an appropriate remediation plans was established. The parameters included proper *P. vittata* ecotype, accompanying microorganisms, chemical emendments. And depending on the local environmental conditions, scenarios for the irrigation, harvest time and frequency, and the disposal method of the As-enriched biomass were to be established.

Often on yearly interval, the pollution status was re-assessed. When the As concentration in soil and the agricultural products meet the national standards, the remediation project is finished. Otherwise, the remediation plan will be remade according to the updated soil pollution assessment result.

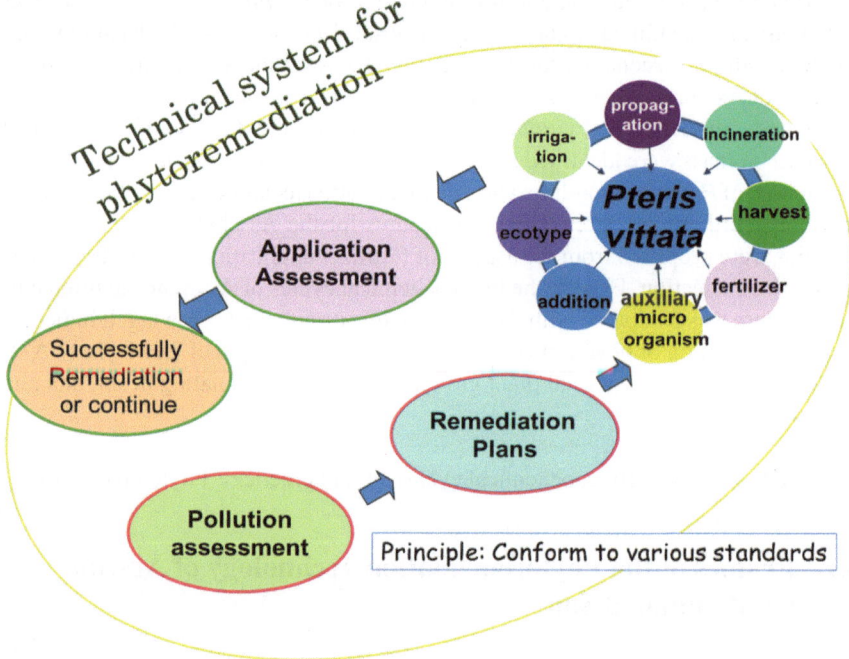

Fig. 3.1 Frames of the phytoremediation technology of arsenic contaminated soil

3.2   Key Process of Phytoremediation Technology of Arsenic Contaminated Soil: Sporelings Culture and Safe Incineration

During these case studies, several money and manpower consuming steps received focused attention.

Fast reproduction of *P. vittata* sporelings. The sufficient supply of *P. vittata* sporelings is the first step for a successful phytoextraction project of As contaminated soil. *P. vittata* is a fern species, which needs spores to propagate. The life cycle of this fern has been described in Chap. 1. The spores of tens of microns can store limited amount of nutrients. After the germination of spores, a sexual fertilization process is necessary for the formation of diploid sporophore. The germination and the fertilization processes have special requirements for light and moisture. Through a series of studies on the germination and the fertilization processes, optimized parameters were proposed (Wan et al. 2010).

With the optimization of sporelings cultivation parameters, the germination rate became much higher (Fig. 3.2), and also the total amount of available sporelings increased with optimizing measures (Fig. 3.3). The main parameters that have been optimized includes soil pH, soil calcium(Ca) content, and temperature.

Higher soil pH and Ca concentration improved the germination of *P. vittata* but did not facilitate the growth of the diploid sporophyte. Spores on Ferralisols soil with the lowest pH and Ca concentration, did not germinate during the 180-day experiment period, while germination on other treatments was observed 14–19 days after sowing.

Soil pH and Ca concentration had different impacts on the germination and later growth of *P. vittata*. On strong acid soils with pH lower than 5, spores of *P. vittata* could not germinate, while on slightly acid soils (pH > 6), germination could occur but the percentage of germination was relatively low; and the appearance time of gametophyte was comparatively late. On alkaline soils, *P. vittata* had higher percentage of germination and earlier appearance time of gametophyte. However, after germination, low pH and Ca concentration facilitated the growth. The period of sporophyte was shortened as the pH of soil decreased. It is suggested that higher soil pH and Ca concentration enhanced germination but inhibited afterward growth, which is conversely improved by lower soil pH and Ca concentration.

Temperature also had a notable impact on the germination and growth afterwards of *P. vittata*. 25 °C was a appropriate temperature for spores to germinate compared to other two temperatures set in this experiment, i.e. 20 and 30 °C. The percentage germination of these three treatments was 25 °C > 20 °C > 30 °C, while Time of gametophyte was 25 °C < 20 °C < 30 °C, being 14.2 d, 25.1 d and 28.4 d, respectively.

25 °C was a appropriate temperature for *P. vittata* to germinate. *P. vittata* raised in environment with a temperature higher or lower than that displayed kind of negative symptoms, such as the small leaf area of gametophyte and the lower maturity rate of gametophyte. Under the temperature of 30 °C, spores had a short period of

Without optimizing measures

With optimizing measures

Fig. 3.2 Comparison of the germination without and with optimizing measures

Fig. 3.3 Increase in sporelings with time. I2, I5 and G1 represents different combinations of optimizing measures, whereas Ck indicated the treatment without any adjustment

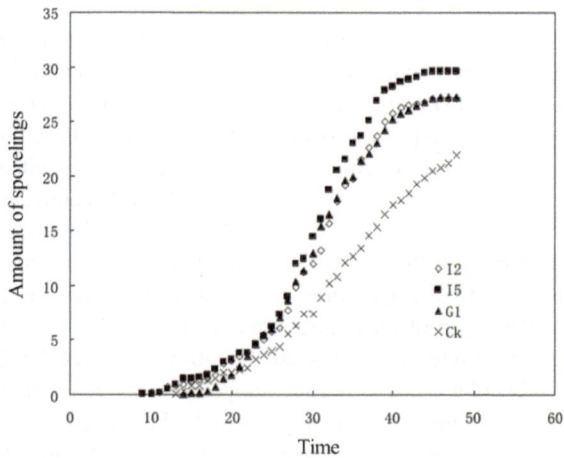

gametophyte but a low percentage of germination and low fertilizing rate. It was also observed that the prothallium diameter was relatively smaller than normal level. Therefore, an extraordinarily high temperature may inhibit the germination of ferns and at the same time inhibit the maturation rate of prothallium by means of influencing the prothallium diameter.

Incineration of hyperaccumulator biomass. The disposal of hyperaccumulators enriched with toxic As is an essential step for a As phytoextraction project. Incineration, direct disposal, ashing and liquid extraction were four predominant methods of hyperaccumulator harvests disposal. Among them, the incineration is proposed as the most feasible, economically acceptable and environmentally sound method (Sas-Nowosielska et al. 2004). Incineration can decrease the volume and the weight of hyperaccumulator biomass by ~90%. During the incineration, the risk caused by the emission of As to the atmosphere needs to be under control.

Disposal of *P. vittata* biomass with extremely high concentration of As is an important step for the phytoextraction practice. Incineration results revealed that 24% of total As accumulated by *P. vittata* containing high As content ($1170 \ \text{mg kg}^{-1}$) is emitted at 800 °C, of which 62.5% of the total emitted As is volatilized below 400 °C. EXAFS and thermo gravimetric experiments showed that carbon originating from biomass incineration might catalyze As(V) reduction (Yan et al. 2008).

Further incineration parameters were optimized by studying the reaction mechanism of As capture by a calcium-based sorbent during the combustion of As-enriched *P. vittata* biomass (Lei et al. 2019). This study investigated three *P. vittata* L. biomass with a disposal capacity of $600 \ \text{kg d}^{-1}$ and different As concentrations from three sites in China (Fig. 3.4).

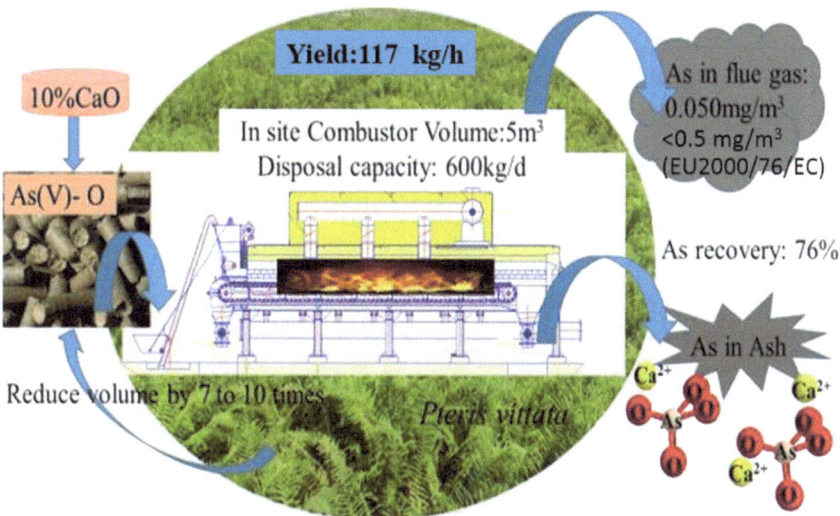

Fig. 3.4 Illustration of the incineration of *P. vittata* biomass (Lei et al. 2019)

   The As concentration in flue gas was greater than that of the national standard in the trial with no emission control, and the As concentration in biomass was 486 mg kg^{-1}. The addition of CaO notably reduced arsenic emission in flue gas, and absorption was efficient when CaO was mixed with biomass at 10% of the total weight. For the trial with 10% CaO addition, arsenic recovery from ash reached 76%, which is an ~8-fold increase compared with the control (Fig. 3.5).

   Synchrotron radiation analysis confirmed that calcium arsenate is the dominant reaction product (Fig. 3.6). This indicates that this Ca-based sorbent can precipitate with As during the incineration process, greatly reduce the released As.

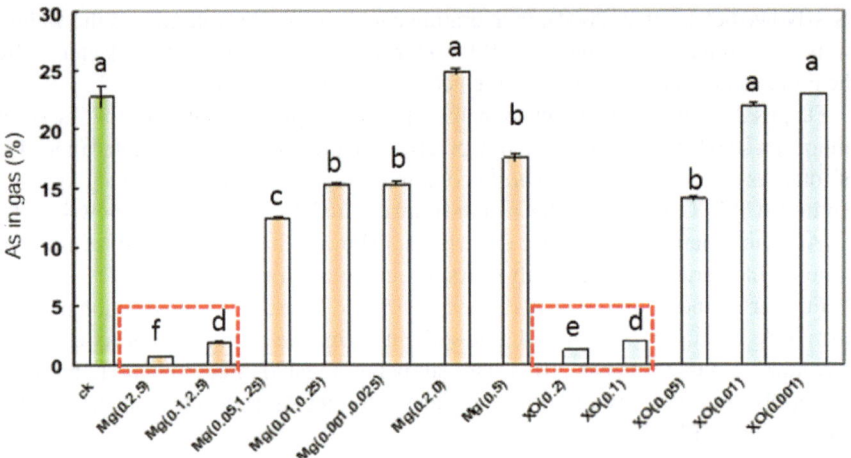

Fig. 3.5 Effects of As fixing agents on the As concentration in gas. Different letters beside the bar indicate significant difference in As in gas among treatments

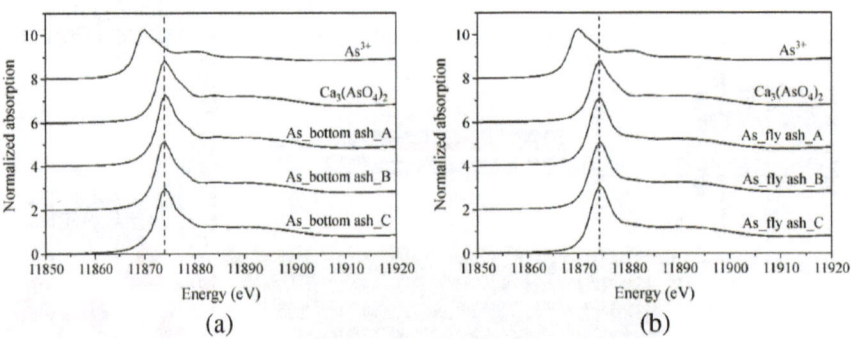

Fig. 3.6 XANES normalized As K-edge profiles in ash samples: **a** Normalized As K-edge profiles for bottom ash samples and reference As compounds. **b** Normalized As K-edge profiles for fly ash samples and reference As compounds. *Note* Bottom and fly ashes A, B, and C are ash samples collected from combustion trials using JY biomass blended with 15%, 10%, and 5% CaO, respectively (Lei et al. 2019)

With the development of the phytoextraction technology of As contaminated soil, the safe disposal of As-enriched biomass was regarded as not sustainable enough. It is required that this biomass should be reused as resources. Agromining was proposed as a method to reuse high-value-metal-enriched biomass of hyperaccumulator, and it has been successfully applied to Ni hyperaccumulator (Echevarria et al. 2015; Kidd et al. 2015; Rosenkranz et al. 2015). However, the potential of As phytomining is not as promising as Ni phytomining, due to the low commercial value of As. Recently, some studies explored a hydrothermal method to convert biomass into biomass fuels with high calorific value (Srokol et al. 2004). In this case, not only valuable Ni was recycled from plants, but also bio-fuel was obtained. Carrier et al. (2011) applied supercritical conditions to safely dispose the biomass of *P. vittata*, they transferred biomass to liquid fuel. At present, scientists from Chinese Academy of Sciences and cooperators from Cranfield University are trying to re-use *P. vittata* biomass as energy. Hopefully the post-process energy and elements recovery from biomass can significantly increase the financial viability of phytoextraction projects and reduce the environment impacts of contaminated biomass disposal (Jiang et al. 2015).

3.3 Strengthening Strategies for Phytoremediation Technology

3.3.1 Fertilizer Application

Aiming to strengthen the strategies for phytoremediation of As contaminated soil using *P. vittata*, the optimum nitrogen (N) and P fertilizers category and amount were selected. The total As accumulation in the plants grown in As-supplied soil, under different forms of N fertilizer, decreased as NH_4HCO_3 > $(NH_4)_2SO_4$ > urea > $Ca(NO_3)_2$ > KNO_3 > CK. The plants treated with N and As accumulated up to 5.3–7.97 mg As pot^{-1} and removed 3.7–5.5% As from the soils, compared to approximately 2.3% of As removal in the control. $NH4^+$-N was apparently more effective than other N fertilizers in stimulating As removal (Liao et al. 2007).

Field experiment showed that the yields of *P. vittata* were enhanced with increasing P addition, however there was no further increase when rate of P addition was more than 200 kg hm^{-2} (Fig. 3.7). Since the interaction between As and P at the plant level has been elaborated in Chap. 2, this chapter only introduce the experiences from the field experiment.

Application of P fertilizer could enhance As concentration within a proper range of dosage, but As concentration in the aboveground of *P. vittata* was depressed because of excessive P addition. Theoretically, As concentration of plant reaches a maximum of 1622 mg kg^{-1} when rate of P addition is 340 kg hm^{-2}. After 7 months of the experiment, soil As concentrations were significantly reduced in all treatments at harvest compared to those before transplanting. When rate of P addition was 200 kg hm^{-2}, efficiency of As removal was the highest (7184%), while efficiencies of As removal

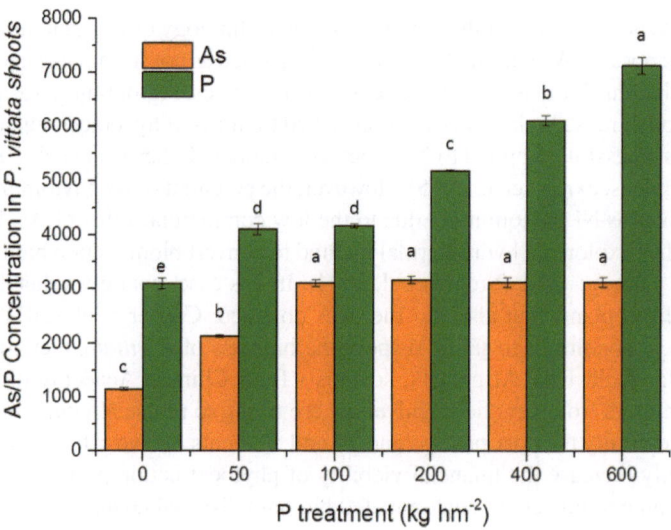

Fig. 3.7 Effect of P treatments on As and P concentrations in shoot of *P. vittata*. Different letters beside the bar indicate significant difference in As or P concentration in *P. vittata* shoots

at control and 600 kg hm^{-2} treatments were 2131 and 6163%, respectively. The highest efficiency of As removal in theory could be achieved at 369 kg hm^{-2} P addition. Moreover, the results also showed that P application was helpful to maintain a balance of available As between before transplanting and after harvest (Liao et al. 2004).

To further identify the optimized P fertilization scenario, another field study was conducted. It indicated that the combined application of N and P fertilizer resulted in an As extraction efficiency of 0.39% (Fig. 3.8).

P fertilizer enhanced plant growth and As accumulation. Removal efficiency of As by different P fertilizer categories were in the order of potassium phosphate (5.9%) > ammonium phosphate (3.5%) > calcium phosphate (1.1%). Removal efficiency of As by different P levels were in the order of 300 kg hm^{-2} (5.9%) > 600 kg hm^{-2} (5.6%) > 150 kg hm^{-2} (1.3%). Potassium phosphate at the level of 300 kg hm^{-2} achieved the highest removal efficiency (28.7%), the proportion of soil total N, P and K is 1: 1.41: 45.5, the As concentration of the soil was reduced from 28 mg kg^{-1} to 19.9 mg kg^{-1}, the treated soil could be safely reused after remediation.

Fertilizer optimization is found to be an efficient strengthening measure to enhance As extraction efficiency and shorten remediation time. And with the development of more advanced and greener fertilizers, the role of fertilizer application in increasing As removal by *P. vittata* might be further enhanced.

Fig. 3.8 Effect of P fertilizer categories and concentrations on As accumulation by *P. vittata*

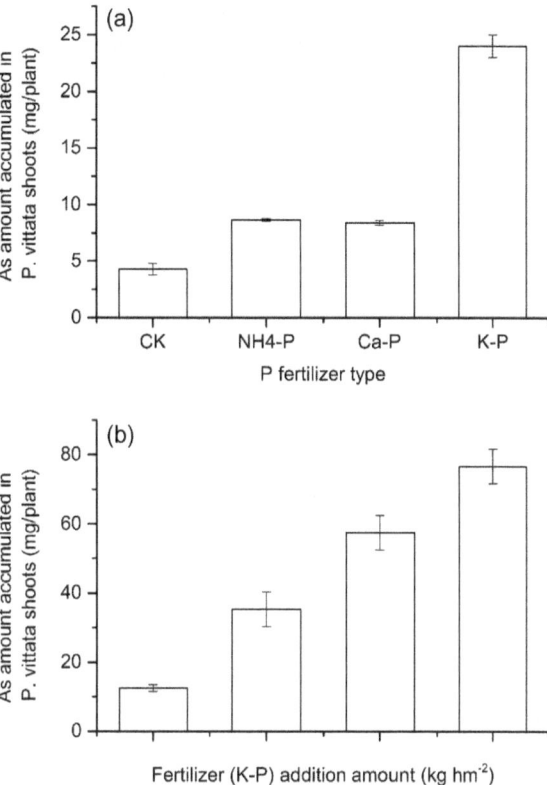

3.3.2 Harvest Cycle

P. vittata is a perennial fern. With the roots kept in soil, the aboveground parts can be harvested for several times a year. For the practical application, the total amount of As accumulation by *P. vittata* is the main indicator to be considered, which depended not only on the As concentration but also the biomass. The effects of harvesting timing and frequency on As accumulation and phytoextraction efficiency of *P. vittata* was studied by both pot and field experiments.

When 0, 100, 300 and 800 mg kg^{-1} of As were added, the total amount of As accumulation by the shoots of *P. vittata* were 216, 1015, 1314 and 1211 mg pot^{-1}, respectively (Fig. 3.9).

The As uptake rates by *P. vittata* in the first harvesting ranged from 20 to 35 μg plant^{-1}, which was significantly lower than those of the second and third harvesting. Therefore, it showed that more harvesting would increase the As accumulation and phytoextraction efficiency of As of *P. vittata*, indicating harvesting was an economical and convenient way for the phytoremediation when utilized *P. vittata* to improve the phytoextraction efficiency of As (Li et al. 2005).

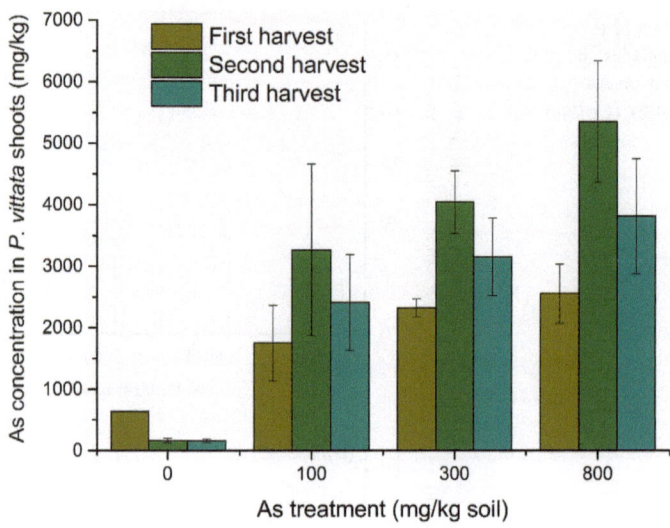

Fig. 3.9 Effect of harvest on As concentrations in shoot of *P. vittata* added with different amounts of As

The results showed that more harvesting would increase the As accumulation and the phytoextraction efficiency of As of *P. vittata*. Through the field practice, 16 kg As, 8.5 kg Pb and 9.2 kg Zn can be removed from one hectare soil each year by harvesting 2 times per year (Xie et al. 2010).

Furthermore, field experiment disclosed the temporal dynamics of biomass and As accumulation by *P. vittata* through sampling after different growth time (Fig. 3.10).

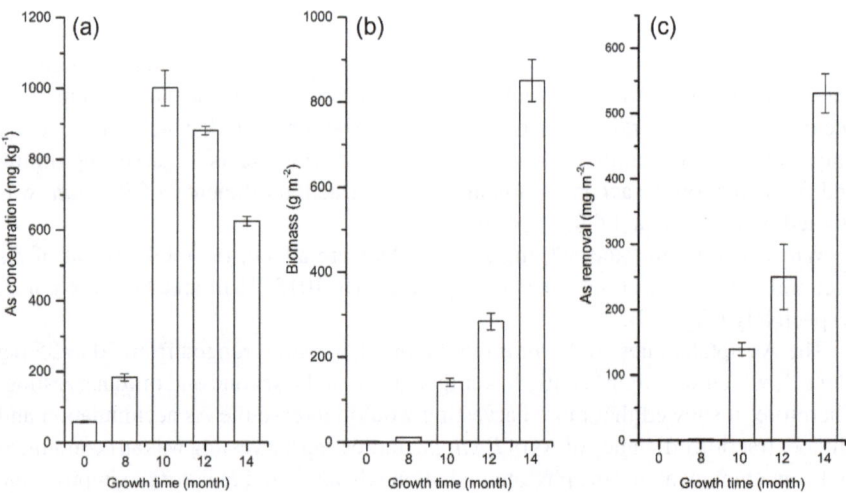

Fig. 3.10 The dynamics of the growth and As accumulation by *P. vittata*

The results showed that the As concentration in *P. vittata* shoots increased in the first several 10 months but decreased in the following 4 months, while the biomass and As amount in *P. vittata* shoots increased with an increase in the growth time. Therefore, it is proposed that the aboveground parts can be harvested 10 months after the initial transplantation, when the aboveground As concentration is the highest.

Optimizing the harvest timing and frequency is another efficient strengthening measure to enhance As extraction efficiency and shorten remediation time. It is noteworthy that this optimization needs to be in line with the local environmental conditions, especially the climate conditions and the available harvest machine.

3.3.3 Ecotype Selection

P. vittata has a wide distribution in China. Through field investigation and lab experiments, it has been found that the variation in As accumulation among different *P. vittata* ecotypes can be ~5.4 times (Wu et al. 2009). Both differences in plant biomass and As accumulating ability among different *P. vittata* ecotypes have been observed (Cai et al. 2007).

Chapter 2 has illustrated the significant difference in As accumulation by different *P. vittata* ecotypes. Therefore, in this chapter, only the application of ecotype selection as a strengthening strategy was introduced.

There were dramatic difference in plant height, biomass and amount of fronds. The plant height ranged from 2916 to 6812 cm, frond number from 1810 to 6010 per plant, shoot fresh weight from 150 to 540 g per plant and root fresh weight from 2013 to 9419 g per plant. The distinct difference among the ecotypes was also found in shoot and root As accumulation, ranging from 643.10 to 3009.03 mg kg^{-1} and from 26.34 to 112.38 mg kg^{-1}, respectively. There was a positive association between As accumulation and some morphology characteristics, including plant height, frond and bud numbers, and shoot fresh weight.

Due to such huge difference in As uptake among *P. vittata* ecotypes, ecotype selection has become an important strengthening strategy to improve As removal efficiency during the remediation practice. The utilization of various ecotypes led to a 3.5-fold difference in As removal rate (Fig. 3.11). This indicates that ecotype selection is a feasible and easy way to enhance the remediation of As contaminated soil.

Not only the different As accumulation capacities were utilized to improve the remediation efficiency, the possibility of using multi-elements accumulating ability to simultaneously extract multi-elements was investigated (Fig. 3.12). There are certain differences in the As and Pb accumulation capacity of the 4 *P. vittata* ecotypes. The 20H ecotype has significantly higher As concentration in the aboveground parts than other ecotypes and significantly lower root As concentration, indicating a strong As

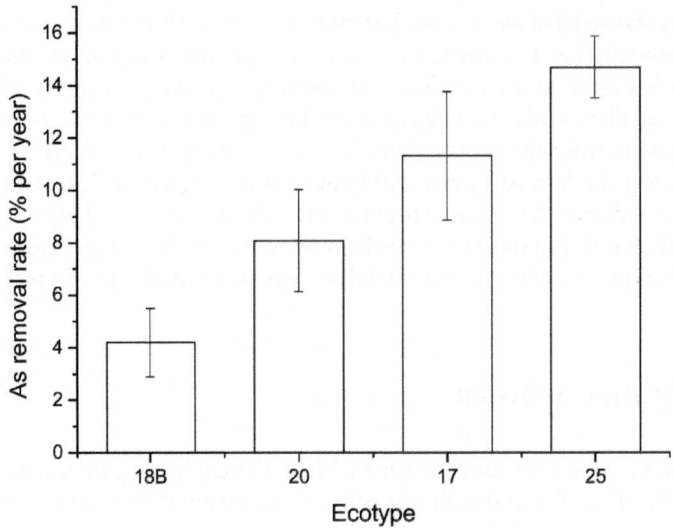

Fig. 3.11 Effect of ecotype selection on the As removal rate

Fig. 3.12 Effect of ecotype
selection on the co-removal
of As and Pb

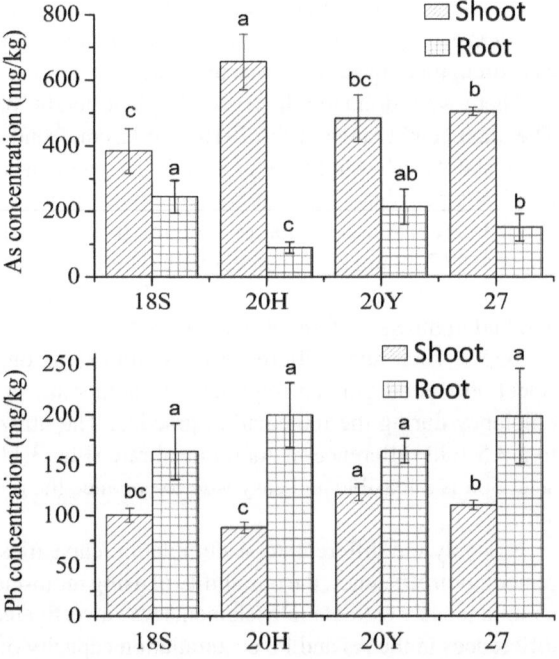

Fig. 3.13 Effect of ecotype selection on the soil As and Pb concentration

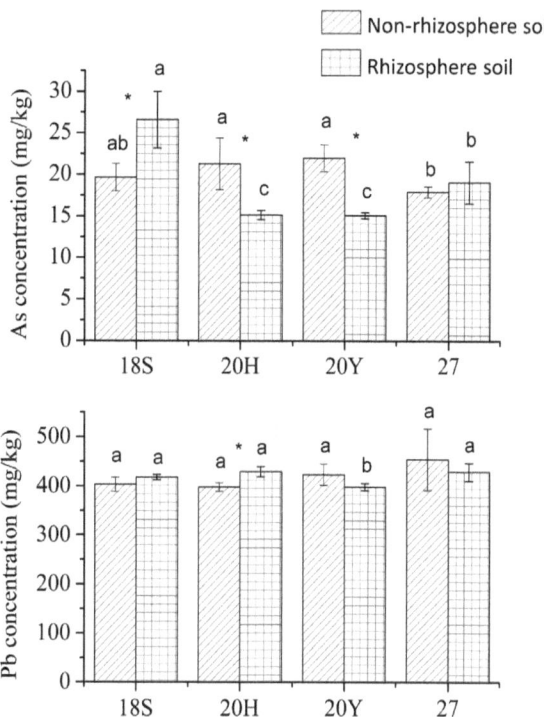

transport capacity. Ecotype 20H and ecotype 27 have a relatively high Pb concentration in the aboveground part, and ecotype 20H also has a relatively low Pb concentration in the root, which indicates that ecotype 20H not only has a strong As transport capacity, but also has a strong Pb transport capacity.

The results of As and Pb concentrations in rhizosphere soil and non-rhizosphere soil showed significant difference in As concentrations between rhizosphere soil and non-rhizosphere soil of different *P. vittata* ecotypes (Fig. 3.13). The As concentration in rhizosphere soil of 18S ecotype was significantly higher than that in non-rhizosphere soil, while that in 20H and 20Y ecotype was significantly lower than that in non- rhizosphere soil, and there was no significant difference in 27 ecotype.

The results showed that the uptake rate of As in roots of 20 h and 20Y ecotypes led to the symptom of arsenic depletion in rhizosphere soil, which was consistent with the higher concentration of As in the aboveground part of the two ecotypes. The difference of Pb concentration between rhizosphere soil and non-rhizosphere soil of different ecotypes is small, only the Pb concentration in rhizosphere soil of ecotype 20H is significantly higher than that in non rhizosphere soil, which may be related to the mobilization of Pb by root exudates.

The selection of appropriate *P. vittata* ecotype is found to be an easy and cheap but very efficient strengthening measure to enhance As extraction efficiency and shorten remediation time. Through the nearly 20-year collection of various *P. vittata* ecotypes, ecotypes with different specialties have been identified and utilized to the field practice of As-contaminated soil phytoremediation.

3.4 Biogeochemical Cycles of Arsenic in the Phytoremediation Field

The schematic diagram of As cycling in the As-contaminated soil phytoremediation system is shown in Fig. 3.14. The input of As into soil includes fertilizer, irrigation, atmospheric deposition, leaves leaching, defoliation. The output includes removal of As by plants, underground leaching, evaporation and surface runoff (Table 3.1).

The input and output of As was calculated for the system with normal plants growing and that with *P. vittata* planted. The As absorbed by normal plants is only limited to the non-specific adsorbed As in the rhizosphere, while *P. vittata* can absorb the non-specific adsorbed arsenic in the rhizosphere and non rhizosphere at the same

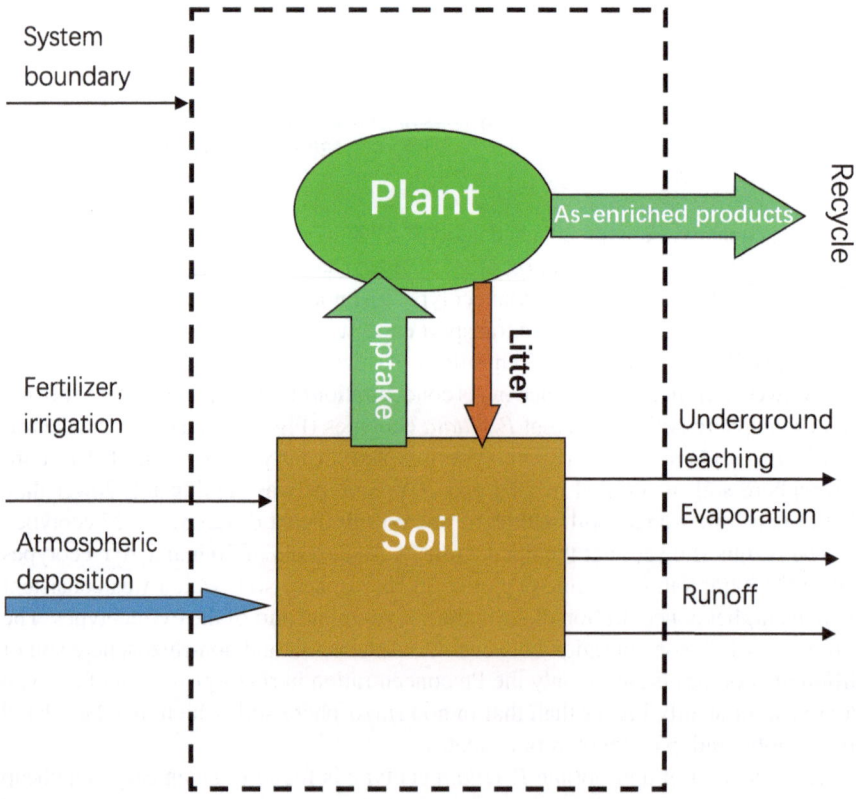

Fig. 3.14 Schematic diagram of As cycling in phytoremediation system

Table 3.1 Input and output of As in the normal plant growing system and *P. vittata* growing system

	P. vittata	Normal plant
Atmospheric deposition	+0.37	+0.37
Irrigation and fertigation	+0.025	+0
Leaves leaching	+0.53	+0
Defoliation	+0.02	+0.00
Runoff	−0.01	−0.01
Soil infiltration	−0.05	−0.05
Aboveground parts removal	−7.90	−0.07
Net influx	−7.02	+0.24

time, and even the specific adsorbed arsenic that is regarded as non-mobile. The input and output fluxes such as soil leaching, surface runoff, atmospheric deposition and plant uptake were monitored annually (Table 3.1).

Based on the annual monitoring data, the As cycling process and dynamic change characteristics in the phytoremediation process of As contaminated soil were disclosed. This establishes the environmental biogeochemical model of As in the phytoremediation system.

Compared with the polluted farmland ecosystem without remediation, As hyper-accumulator can significantly change the As cycling behavior in the soil ecosystem (Fig. 3.15). In the phytoremediation process, the main As input pathway of soil system is atmospheric deposition, with the annual input flux being 0.94 kg ha^{-1} A^{-1}, while the annual output fluxes of As 7.96 kg ha^{-1} A^{-1}. For the phytoreme-diation system with *P. vittata* growing, the overall As cycling performance is net output, being 7.02 kg ha^{-1} A^{-1}. Whereas for the normal plant growing system, the

Fig. 3.15 Net input and output of As in the normal plant growing system and *P. vittata* growing system

net input of As is 0.24 kg ha^{-1} A^{-1}. The As clean-up ratio by hyperaccumulators was quantified.

Through a series of field investigation, lab experiment, and field practice, the phytoremediation technology frame was established. And the key process was identified and optimized. Furthermore, the strengthening strategies for phytoremediation technology were proposed. Finally, the biogeochemical cycles of As in the phytoremediation system was compared with the system with normal plants growing. It indicated that the application of phytoremediation led to an As net output of 7.02 kg ha^{-1} A^{-1}, while on the field with normal plant growing, As was accumulating in soil at a rate of 0.24 kg ha^{-1} A^{-1}.

References

Cai B, Zhang G, Chen T (2007) Genotypic difference in growth and As accumulation in *Pteris vittata*. J Zhejiang Univ 33:473–478

Carrier M, Loppinet-Serani A, Absalon C, Marias F, Aymonier C, Mench M (2011) Conversion of fern (*Pteris vittata* L.) biomass from a phytoremediation trial in sub- and supercritical water conditions. Biomass Bioenerg 35:872–883

Chen TB, Wei CY, Huang ZC, Huang QF, Lu QG, Fan ZL (2002) Arsenic hyperaccumulator *Pteris vittata* L. and its arsenic accumulation. Chin Sci Bull 47:902–905

Ebbs S, Hatfield S, Nagarajan V, Blaylock M (2010) A Comparison of the dietary arsenic exposures from ingestion of contaminated soil and hyperaccumulating *pteris* ferns used in a residential phytoremediation project. Int J Phytorem 12:121–132

Echevarria G, Baker A, Benizri E, Morel JL, Van Der Ent A, Houzelot V, Laubie B, Pons MN, Simonnot MO, Zhang X, Kidd P, Bani A (2015) Agromining for Nickel: a complete chain that optimizes ecosystem services rendered by ultramafic landscapes

Eisler R (1985) A review of arsenic hazards to plants and animals with emphasis on fishery and wildlife resources. In: Nriagu JO (ed) Arsenic in the environment, part I: cycling and characterization. Wiley, New York (USA), pp 185–259

Huang Z-C, An Z-Z, Chen T-B, Lei M, Xiao X-Y, Liao X-Y (2007) Arsenic uptake and transport of *Pteris vittata* L. as influenced by phosphate and inorganic arsenic species under sand culture. J Environ Sci-China 19:714–718

Jiang Y, Lei M, Duan L, Longhurst P (2015) Integrating phytoremediation with biomass valorisation and critical element recovery: a UK contaminated land perspective. Biomass Bioenerg 83:328–339

Kidd P, Alvarez-Lopez V, Quintela-Sabaris C, Cabello-conejo MI, Prieto-Fernandez A, Becerra-Castro C, Monterroso C (2015) Improving the Nickel Phytomining capacity of hyperaccumulating subspecies of *Alyssum serpyllifolium*

Lei M, Dong Z, Jiang Y, Longhurst P, Wan X, Zhou G (2019) Reaction mechanism of arsenic capture by a calcium-based sorbent during the combustion of arsenic-contaminated biomass: a pilot-scale experience. Front Environ Sci Eng 13

Li W, Chen T, Liu Y (2005) Effects of harvesting on As accumulation and removal efficiency of As by Chinese brake (*Pteris vittata* L.) (in Chinese, and abstract in English). Acta Ecol Sin 25:538–542

Liao X-Y, Chen T-B, Xiao X-Y, Xie H, Yan X-L, Zhai L-M, Wu B (2007) Selecting appropriate forms of nitrogen fertilizer to enhance soil arsenic removal by *Pteris vittata*: a new approach in phytoremediation. Int J Phytorem 9:269–280

Liao X, Chen T, Xie H, Xiao X (2004) Effect of application of P fertilizer on efficiency of As removal form As contaminated soil using phytoremediation: field study (in Chinese, and abstract in English). Acta Sci Circum 24:455–462

Rosenkranz T, Puschenreiter M, Kisser J (2015) Phytomining of valuable metals from waste incineration bottom ash using hyperaccumulator plants

Sas-Nowosielska A, Kucharski R, Malkowski E, Pogrzeba M, Kuperberg JM, Krynski K (2004) Phytoextraction crop disposal—an unsolved problem. Environ Pollut 128:373–379

Sirkin HL, Zinser M, Hohner D (2011) Made in America, again: why manufacturing will return to the US

Srokol Z, Bouche AG, van Estrik A, Strik RCJ, Maschmeyer T, Peters JA (2004) Hydrothermal upgrading of biomass to biofuel; studies on some monosaccharide model compounds. Carbohyd Res 339:1717–1726

Sturchio E, Boccia P, Meconi C, Zanellato M, Marconi S, Beni C, Aromolo R, Ciampa A, Diana G, Valentini M (2013) Effects of arsenic on soil-plant systems. Chem Ecol 27:67–78

Wan X-M, Lei M, Huang Z-C, Chen T-B, Liu Y-R (2010) Sexual propagation of *Pteris vittata* L. Influenced by pH, Calcium, and temperature. Int J Phytorem 12:85–95

Wan X, Lei M, Chen T (2016) Cost–benefit calculation of phytoremediation technology for heavy-metal-contaminated soil. Sci Total Environ 563–564:796–802

Wu FY, Leung HM, Wu SC, Ye ZH, Wong MH (2009) Variation in arsenic, lead and zinc tolerance and accumulation in six populations of *Pteris vittata* L. from China. Environ Pollut 157:2394–2404

Xie J, Lei M, Chen T, Li X, Gu M, Liu X (2010) Phytoremediation of soil co-contaminated with arsenic, lead, zinc and copper using *Pteris vittata* L.: a field study (in Chinese, abstract in English). Acta Sci Circum 30:165–171

Yan X-L, Chen T-B, Liao X-Y, Huang Z-C, Pan J-R, Hu T-D, Nie C-J, Xie H (2008) Arsenic transformation and volatilization during incineration of the hyperaccumulator *Pteris vittata* L. Environ Sci Technol 42:1479–1484

Chapter 4
Application of Phytoremediation Technology to Typical Mining Sites in China

Abstract In situ phytoextraction projects using *Pteris vittata* have been established with high removal rates of As. The first phytoextraction project in the world was established in Chenzhou of Hunan Province. Afterwards, more phytoextraction projects were established in Guangxi Zhuang Autonomous Region, Yunnan Province, and Henan Province. Through the phytoremediation practices, phytoremediation is confirmed to be an economical, easy and green technology, which well fit for the current farmland contamination status. Besides, the strengthening measures have expanded the application range of phytoextraction from south China to North China. And the utilization of this technology is not limited to farmland but also works for the brown sites. However, there are still a lot of questions waiting for further study. These questions proposed new demands for the science and technology development of phytoremediation technology.

Keywords Agricultural land · Application · Field · Mining site

4.1 In Situ Phytoremediation Projects by *P. vittata*

In situ phytoremediation projects using *P. vittata* have been established on both farmlands in China and residential areas in America, with high As removal rates achieved (Chen et al. 2007; Ebbs et al. 2010). The first phytoextraction project in the world was established in Chenzhou, Hunan Province. After the cultivation of *P. vittata* for seven months, As concentration in the soil decreased by 5.0 mg kg^{-1}, with remediation efficiency reaching 7.84% (Liao et al. 2004). Afterwards more phytoextraction projects were established in the Guangxi Zhuang Autonomous Region, Yunnan Province, Henan Province, Guangdong Province, and Hebei Province (Chen et al. 2018a, b). In these projects, the main targeted contaminant is As, accompanied by cadmium (Cd) and lead (Pb) under most conditions.

Table 4.1 summarizes the representative phytoremediation projects of arsenic contaminated soil in China. The main technologies are phytoextraction and intercrop. The As removal rate reached ~18%.

© The Author(s) 2020
T. Chen et al., *Phytoremediation of Arsenic Contaminated Sites in China*,
SpringerBriefs in Environmental Science,
https://doi.org/10.1007/978-981-15-7820-5_4

Table 4.1 Phytoremediation projects of arsenic contaminated soil in China

Place	Contaminant	Technology	Remediation efficiency
Chenzhou, Hunan Province	As	Phytoextraction	Through 3–5 year remediation, soil As content decreased from 50 mg kg^{-1} to <30 mg kg^{-1}, reaching the second national soil quality standard (GB15618-1995)[a]
Shimen, Hunan Province	As	Phytoextraction and intercropping	Phytoextraction technology removed 13% As from soil each year In the intercropped system, agricultural products from intercropped cash crops meet the national standard
Huanjiang, Guangxi Zhuang Autonomous Region	As, Cd and Pb	Phytoextraction, intercropping and phytobarrier	Phytoextraction technology removed 10.5% Cd and 28.6% As from soil after 2-year remediation In the intercropped system, the yield of maize, rice and sugarcane increased by 154%, 29.6% and 105%, respectively; and the As, Cd and Pb concentration in corn kernel decreased by 39%, 4.1 and 4.9%, respectively In the phytobarrier system, the over standard rate of heavy metals in agricultural products was less than 5%
Gejiu, Yunnan Province	As and Pb	Phytoextraction	Phytoextraction technology removed 18% As and 14% Pb from soil each year
Huize, Yunnan Province	As	Phytoextraction	Phytoextraction technology removed 12% As from soil each year
Jiyuan, Henan Province	As, Cd and Pb	Phytoextraction	Phytoextraction technology removed 13.9% As, 0.5% Pb and 16.1% Cd from soil each year After 2-year remediation, 338.5 g As, 36.2 g Cd and 104.5 g Pb were removed from each Mu soil
Fangshan, Beijing City	As	Phytoextraction	Phytoextraction technology removed 17.2% As from soil each year

(continued)

Table 4.1   (continued)

Place	Contaminant	Technology	Remediation efficiency
Dabaoshan, Guangdong Province	As, Cd, Cu, Pb and Zn	Intercropping and phytobarrier	Products of cash crops met the national standards
Yangshuo, Guangxi Zhuang Autonomous Region	As, Cd	Phytoextraction, intercropping of *P. vittata* and Cd accumulating *Amaranthus cruentus* L	Annual removal rate of As and Cd reached 10%, respectively
Baoding, Hebei Province	As	Phytoextraction	Phytoextraction technology removed 13.0% As each year

[a]For the projects finished before 2018, GB15618-1995 was used as the national standard; while for the projects finished after 2018, GB15618-2018 was used as the national standard

4.2   The First Phytoremediation Project for Arsenic Contaminated Soil: Chenzhou, Hunan Province

In 1999, a serious As poisoning incident due to As contamination happened, more than 300 people were hospitalized in Dengjiatang, Chenzhou City, Hunan Province. 50 ha of paddy fields were contaminated and the farmlands were wasted for 4 years, though the source of As contamination had been cut off by local authority in time. Only a few vegetable fields can still be cultivated on the fields. The As concentration of plants and soils on the fields with different levels of As contaminations. The health risk of consumed vegetables cultivated were investigated in the As-contaminated area. The concentration of As in the edible parts of most vegetables were higher than the maximal permissible limit of As in food. The intake of As from consumed vegetables was $4.1\,\mu g\,kg^{-1}$ body wt^{-1} per day in spring and summer, and $2.9\,\mu g\,kg^{-1}$ body wt^{-1} per day in autumn and winter, respectively, both exceeding the maximal allowed level ($2.1\,\mu g\,kg^{-1}$ body wt^{-1} per day) set by World Health Organization (WHO).

In 2001, the first phytoremediation project for As contaminated soil was established in Dengjiatang, Chenzhou City, Hunan Province (Fig. 4.1). After one-year planting of *P. vittata* on this piece of land, the average concentration of As in soil decreased by ~10%. The highest As concentration in *P. vittata* reached 0.8% (w:w). After 5-year remediation, the soil As concentration decreased from 50 mg kg^{-1} to below 30 mg kg^{-1} (Fig. 4.2). The annual As removal rate was 7.8%. And the agricultural products, mainly vegetables, growing on this soil after phytoremediation, meet the related national food standard.

This project is the first phytoremediation project for As contaminated soil. After five-year practice, photoextraction has been confirmed to be an efficient method to remove As from soil. Based on experiences from this project, more projects have been conducted in other areas with a similar As pollution problem.

Fig. 4.1 The first phytoremediation project for As contaminated soil in Chenzhou, Hunan Province

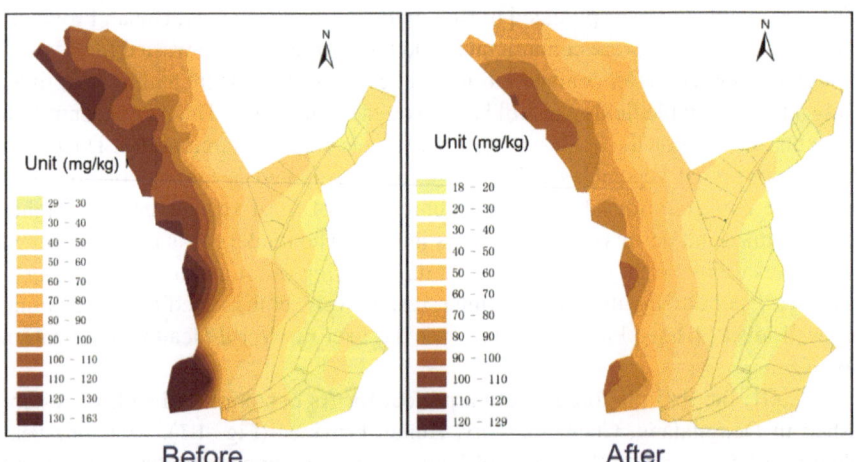

Fig. 4.2 Comparison of As concentration in soil before and after soil remediation

4.3 The Biggest Phytoremediation Project for Arsenic Contaminated Soil: Huangjiang, Guangxi Autonomous Region

The phytoremediation project in Huanjiang, the province of Guangxi was the largest phytoextraction project in the world at that time. Phytoextraction, intercropping and phytobarrier technologies were used. The phytoextraction technology of *P. vittata* reached an annual As removal rate of 10%. The phytobarrier technology significantly decreased As concentration in agricultural products, with less than 5% over-standard rate. The intercropping technology also reached an over-standard rate of As being less than 5%. Through the Huanjiang project, a new model of "government guidance, scientific and technological support, enterprise participation, and the implementation of farmers" was established and the experiences from Huanjiang have been utilized on other sites.

4.3.1 Background Information

Location of Huanjiang County is provided in Fig. 4.3. In the summer of 2001, the Beishan Pb–Zn mine's tailing dam located in the upstream of Huanjiang river collapsed due to a catastrophic flood, leading to the spread of mining waste spills on the farmlands along the River (Fu et al. 2015).

About 700 ha soil were seriously contaminated by heavy-metal-enriched flooding water. Plants can no longer grow on some soils due to serious pollution. The rest soil can produce agricultural products but these products can hardly meet the natural

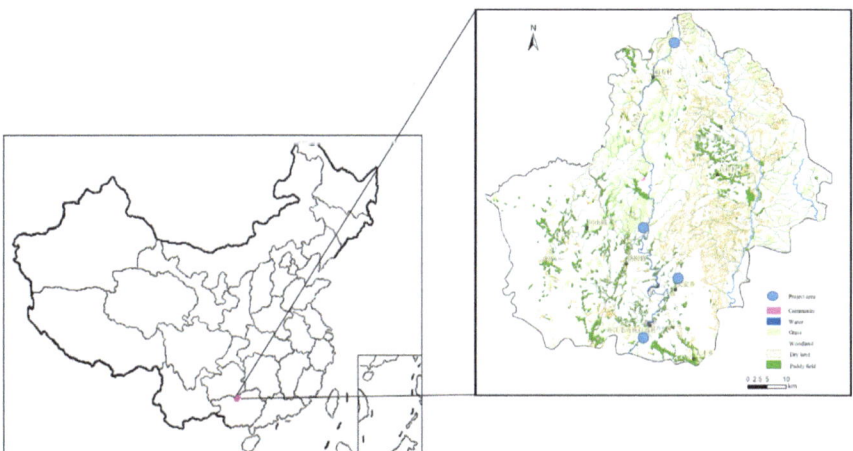

Fig. 4.3 The location of Huanjiang County (Wan et al. 2016)

standards. Local people showed some pathological symptoms after digesting crops produced from contaminated soils, such as the decreased phosphor in plasma, and an increase in toxic elements concentrations in urine. The main contaminants were Cd, Pb and As, with the average concentrations being 0.46 mg kg^{-1}, 350.5 mg kg^{-1} and 44.1 mg kg^{-1}, respectively. The heavy metal contamination in this area has become an impressing environmental issues.

4.3.2   Selection of Remediation Technology

According to the investigation results, the remediation plan was made (Fig. 4.4). For the moderately contaminated soil, phytoextraction technology was adopted. Arsenic and Pb hyperaccumulator *Pteris vittata* (Chen et al. 2002a, b; Ma et al. 2001; Wan et al. 2014) and Cd hyperaccumulator *Sedum alfredii* Hance (Long et al. 2009; Yang et al. 2004) were used there.

To give the local farmers some income when cleaning contaminated soil, inter-cropping system comprised by hyperaccumulators and cash crops with low HMs accumulating abilities (Wang et al. 2015) were established on the slightly contaminated soil. The total area of the remediation project was 19.5 ha, distributing along the Huanjiang river. The phytoextraction area was 11.1 ha, with the inter-cropping of hyperaccumulator and sugar cane being 5.6 ha, and intercropping of hyperaccumulator and mulberry tree being 2.8 ha.

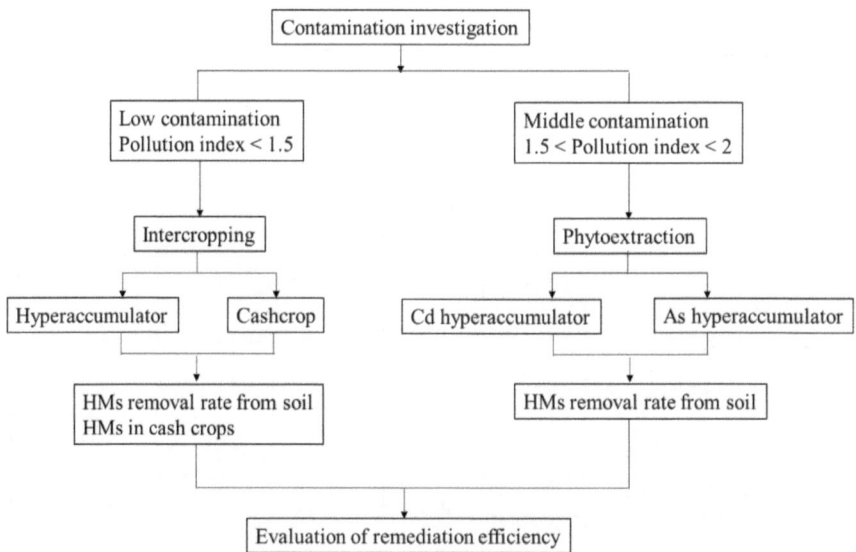

Fig. 4.4 The basic structure of the remediation plan (Wan et al. 2016)

Table 4.2 The heavy metal concentrations in agricultural products after remediation (Wan et al. 2016)

Time	Index	As	Cd	Pb
First year	Average	0.007	0.0106	0.286
n = 51	Median	0	0.005	0.234
Second year	Average	0.014	0.0067	0.167
n = 46	Median	0	0.0048	0.133

4.3.3 Evaluation of Remediation Efficiency

After 2-year remediation, the concentrations of As, Cd and Pb significantly decreased in soil. The concentration of available As decreased by 55.3%, and that of Cd decreased by 85.8%, while that of Pb decreased by 30.4%. Cash crop products produced during the remediation met the national standards and brought in local farmers some financial support. After the remediation, the agricultural products growing on these soils can meet the national standards (Table 4.2).

4.3.4 Calculation of Cost and Benefit

The total cost of phytoremediation was US$75,375.2 hm^{-2} or US$37.7 m^{-3}, with initial capital and operational costs accounting for 46.02% and 53.98%, respectively. The costs of infrastructures (i.e., roads, bridges, and culverts) and fertilizer were the highest, mainly because of slow economic development and serious contamination in the local area. The cost of phytoremediation was lower than the reported values of other remediation technologies (Table 4.3). Improving the mechanization level of phytoremediation and accurately predicting or preventing unforeseen situations were suggested for further cost reduction.

Benefits include benefits during the remediation and that after remediation (Table 4.4). The benefit during remediation mainly comes from the cash crops in the intercropping system, including 5.6 ha of sugar cane and 2.8 ha of mulberry tree. Sugar cane and mulberry tree produced an income of 90,932.8 and 45,220 U.S. dollar during the two-year remediation, respectively. Converting this value to an average benefit per hectare soil, it was 4663.2 and 2318.9 U.S. dollar per hectare soil. The benefit after the remediation includes the benefit of recovering the function of soil to produce agricultural products (B_{AP}), the benefit of recovering soil as a healthy ecosystem component (B_{EP}) and the benefit of preventing human income loss (B_{RC}). B_{EP} and B_{RC} were just-for-once benefit, while B_{AP} was a yearly benefit.

Taking both the costs and benefits into consideration, the costs minus the just-for-once benefits equals 55,758.9. It was hypothesized that the inflation rate equals the discount rate. The average benefit of agriculture producing was 8241 U.S. dollar

Table 4.3 The costs of Huanjiang phytoremediation project (Wan et al. 2016)

Items			Costs (U. S. dollar hm^{-2})	Percentage (%)
Initial capitalization	Pollution survey		824.8	1.09
	Establishment of the remediation strategy		824.8	1.09
	Land preparation		577.3	0.77
	Nursery equipment		5893.6	7.82
	Irrigation system		5986.8	7.94
	Roads, bridges and culverts		9548.4	12.67
	Incineration equipment		7216.5	9.57
	Others		3812.4	5.06
Initial capitalization in total			34,684.5	46.02
Operational cost (two years)	Cost of labor	Seedling	164.9	0.22
		Plough	780.8	1.04
		Transplant	206.1	0.27
		Fertilize	123.7	0.16
		Insect control	123.7	0.16
		Irrigation	123.7	0.16
		Weed control	412.4	0.55
		Harvest	185.6	0.25
		Others	657.8	0.87
		Cost of labor in total	2778.7	3.69
	Cost of materials	Seedling tray	82.5	0.11
		Hyperaccumulator seedlings	164.9	0.22
		Crops seedlings	2521.6	3.35
		Farm chemicals	41.2	0.05
		Fertilizer	14,891.9	19.75
		Others	922.7	1.22
		Cost of materials in total	18,624.9	24.71
	Cost to use large machines	Harvest machine	296.9	0.39
		Incineration machine	321.6	0.43
		Disposal of dangerous wastes	206.1	0.27
		Cost to use machines in total	824.8	1.09

(continued)

Table 4.3 (continued)

Items			Costs (U. S. dollar hm^{-2})	Percentage (%)
	Other direct cost	Production compensation	356.7	0.47
		Rent of land	309.3	0.41
		Fuel and power cost	1948.4	2.58
		Construction supervision	74.2	0.10
		Environment supervision	4011.2	5.32
		Regular monitor	3299	4.38
		Other direct cost in total	9998.8	13.27
	Indirect cost	Staff wage	989.7	1.31
		Administrative expenses	824.8	1.09
		Travel expenses	3888.6	5.16
		Cost of water and electricity	2005.8	2.66
		Others	754.7	1.00
		Indirect cost in total	8463.6	11.23
Operational costs in total			40,690.7	53.98
Costs in total			75,375.2	100

Table 4.4 The benefits of Huanjiang phytoremediation project (Wan et al. 2016)

Items		U.S.dollar hm^{-2}
Benefits during remediation (B_{DR})	Sugar cane	4663.2
	Mulberry tree	2319.0
Benefits after remediation (B_{AR})	Ecosystem service function (B_{EP})	1015
	Decrease in human income loss (B_{RC})	11,619.1
	Agricultural products producing function per year (B_{AP})	8241.0 per year

B_{DR}, B_{EP} and B_{RC} were just-for-once benefits, while B_{AP} was a yearly benefit

hm^{-2}, then in less than 7 years, the benefits would offset the costs used for the phytoremediation.

However, it should be noticed that the calculation of the benefit of recovering soil as a healthy ecosystem component and the benefit of preventing human income loss might be more like a semi-quantitative calculation. Further establishment of accurate quantification method is necessary to better evaluate the cost-benefit balance.

The phytoremediation project in Huanjiang showed that this technology can decreased the concentrations of As, Cd and Pb in soil to the level of below the national standards. The total cost of phytoremediation was 75,375.2 U.S. dollar hm^{-2}, or 37.7 U.S.dollar m^{-3}, lower than most technologies reported in the literature. Improving the mechanization level of phytoremediation and accurately predicting or preventing the unexpected were suggested to be able to further decrease the cost. The benefits of phytoremediation project were expected to offset the phytoremediation costs in less than seven years.

4.4   The First Enterprise-Paid as Contaminated Brown Site Remediation Project: Yunnan Chihong Zn & Ge Co., LTD

Yunnan Chihong Zn & Ge Co., LTD has a large production of Pb and Zn ore related products. During the mining and smelting processes, sites contaminated by As, Pb, Cd, Zn and Cu appeared (Table 4.5). According to the investigation, the order of the pollution degree for each pollutant was As > Pb > Cd > Zn > Cu. The spatial distribution of As, Pb, Zn, Cu, Ni and Fe was very similar, indicating their same pollution source.

Phytoextraction using hyperaccumulator *P. vittata* was conducted there (Fig. 4.5). The Pb concentration in *P. vittata* reached 1303 mg kg^{-1} in fronds, meeting the requirement of Pb hyperaccumulator. *P. vittata* can remove 15.5 kg ha^{-1} As and 8.5 kg ha^{-1} Pb from soil per year (2 harvests). When the concentrations of As and Pb in soil was less than 1058 mg kg^{-1} and 5660 mg kg^{-1}, respectively, *P. vittata* displayed higher potential removing As and Pb. In addition, *P. vittata* showed strong tolerance to As, Pb, Zn. The biomass of *P. vittata* was not significantly decreased when the soil concentrations of Pb, Zn and Cu were as high as 10,913 mg kg^{-1}, 2511 mg kg^{-1} and 510 mg kg^{-1}, respectively. After 3-year remediation, the concentration of As in soil decreased by 18% while the concentration of Pb in soil decreased by 14%.

This project was the first enterprise-paid As contaminated brown site remediation project in China, which indicated that phytoremediation can not only be used to agricultural soil but also to contaminated sites. In addition, it indicated that the

Table 4.5 The heavy metal concentration of soil from the contaminated site

Index	Heavy metal				
	As	Cr	Cu	Pb	Zn
Min.	8.13	1.99	3.93	4.37	125
Max.	843	363	1718	15,572	13,750

Fig. 4.5 Photo of the project site during remediation

phytoremediation technology may be not only used to government-led projects but also enterprises-led projects.

4.5   Expand the Arsenic Phytoremediation Practice to Northern China: Jiyuan Henan Province

It has been mentioned in the fist chapter that the natural distribution of *P. vittata* was mainly in south China because the spores can only germinate when the average temperature in January (the coldest month in China) was below 0 °C (Wan et al. 2010). However, As-contaminated soil is distributed not only in Southern China where mining activities are active but also in Northern China where irrigation of wastewater and mine production frequently occur (Gao et al. 2015; Heimann et al. 2015; Li et al. 2015; Tang et al. 2015). Low temperature during winter might be a limiting factor of the application of *P. vittata* in Northern China. It has been introduced in the first Chapter that although *P. vittata* cannot germinate under 20 °C but it has been found that after germination, the sporophyte might be able to grow under low temperature. The possibility of using *P. vittata* to remediate As-contaminated soil in Northern China with appropriate cover materials was investigated.

Jiyuan, Henan Province has a temperate continental monsoon climate featuring four distinct seasons. Due to the low temperature in winter, there is no natural distribution of *P. vittata* in Jiyuan. Henan Yuguang Au & Pb Co., located in the suburb of Jiyuan city, is the biggest Pb smelter company in China. The soil around Yuguang Au & Pb Co. has been contaminated, with main contaminants being As, Pb, and Cd. The remediation site, with an area of approximately 10,000 m^2, is in the southwest of Yuguang Au & Pb Co. (Fig. 4.6), which is mainly contaminated by As and Pb.

The average concentrations of As and Pb were 25.2 and 259.2 mg kg^{-1}, respectively. During the initial investigation, 33 and 100% of the collected samples showed an As and Pb concentration above the maximal allowed concentrations of As (0.5 mg kg^{-1}) and Pb (0.2 mg kg^{-1}) in food (GB 2762-2012).

The distribution of As and Pb showed similar pattern: As concentration in northeast was higher that that in the west. In accordance with the contamination trends of soil, there were more samples indicating a Pb concentration higher than the national food standard than As concentration. 82.6% of the collected samples indicated higher Pb concentration than the maximum limit regulated in the national standard. Whereas only 2 out of 32 samples showed an As concentration higher than the national standard (Fig. 4.7).

Phytoremediation using the As and Pb co-accumulating ecotypes of *P. vittata* was conducted in Jiyuan, Henan Province (Zhang et al. 2016). At the beginning of November, when temperature at night began to drop below 0 °C, all the aboveground biomass of *P. vittata* were removed, leaving only ~5 cm stubble. Different kinds of cover materials were placed on *P. vittata*.

The survival rate and coverage were in the sequence of blank < fabric < film < soil + film < maize straw + film (Fig. 4.8). In the treatment with no cover material,

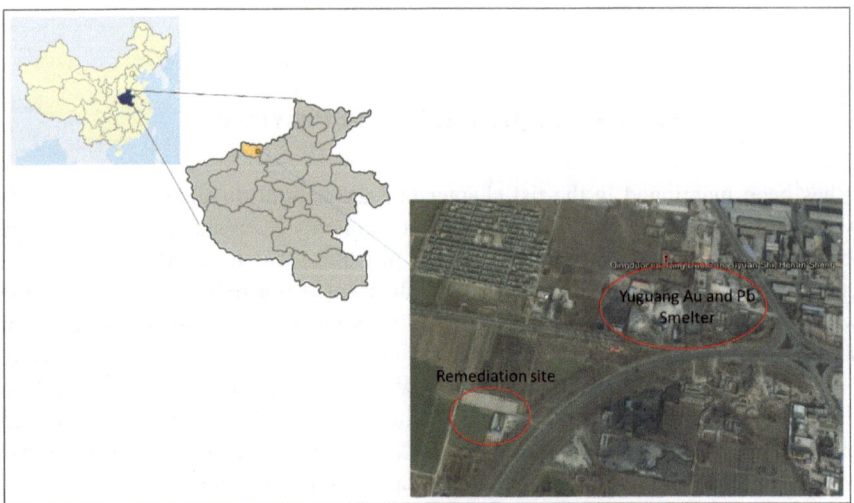

Fig. 4.6 Location of the remediation site in Jiyuan, Henan Province

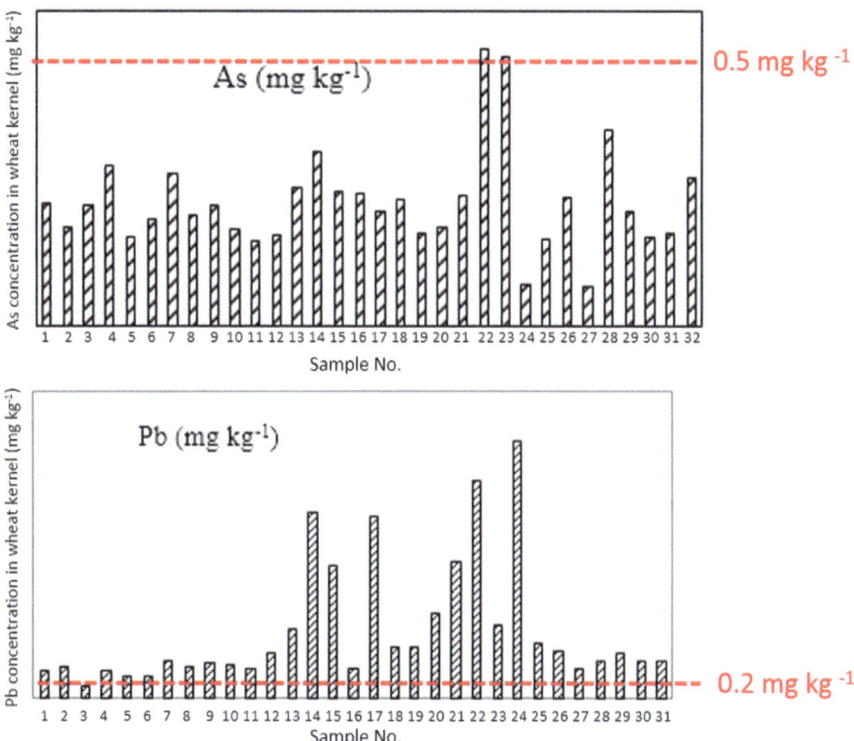

Fig. 4.7   The concentration of As and Pb in wheat kernel. The red dotted line is the maximal allowed concentrations of As (0.5 mg kg^{-1}) and Pb (0.2 mg kg^{-1}) in food (GB 2762-2012)

no *P. vittata* survived, which indicates that without aiding measures, *P. vittata* cannot survive in the winter of Northern China. The treatment of maize straw + film showed the highest survival rate and coverage, being 31% and 47%, respectively. The survival rate and coverage of the treatment of soil + film were slightly lower, being 22% and 45%, respectively. The survival rate and coverage of the treatment of film were lower than their combination, being 18% and 44%, respectively. The treatment of fabric displayed the lowest survival rate and coverage, being 18% and 30%, respectively.

Based on the survival rate and coverage of these four kinds of cover materials, the combination of film and soil or the combination of film and maize straw is a better cover material for *P. vittata* in Jiyuan.

The biomass of individual *P. vittata* under different treatments was not significantly different (Fig. 4.9). The treatment of fabric and single layer of film had a slightly higher individual biomass than the other two treatments, being ~42.2 g plant^{-1} and ~41.6 g plant^{-1}, respectively. Similarly, dry biomass was also slightly higher in the treatment of fabric and film, being 11.8 g plant^{-1} and 10.6 g plant^{-1}, respectively.

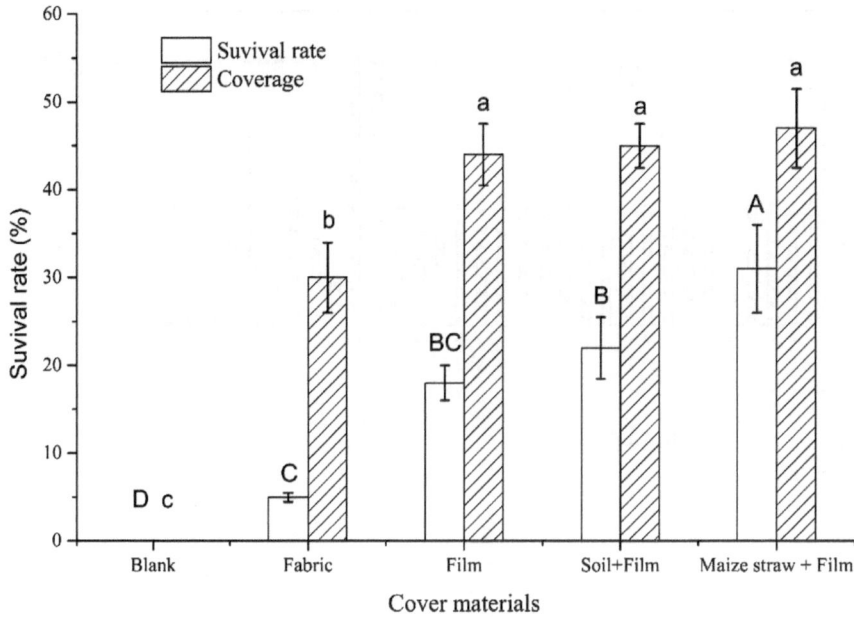

Fig. 4.8 Survival rate and coverage of *P. vittata* with different cover materials. Different capital letters indicate significant differences in survival rate among different treatments; different lowercase letters indicate significant differences in coverage among different treatments ($P < 0.05$)

The treatment of film showed the highest As concentration in the aboveground parts of *P. vittata* (Fig. 4.10). In July, As concentration in *P. vittata* ranged from 209 to 321 mg kg^{-1}. The order of As concentration was film > fabric > soil + film > maize straw + film. The treatment of film had significantly higher As concentration than in the treatment of maize straw + film.

In September, As concentration apparently increased to 460–795 mg kg^{-1}. The order of As concentration changed to film > soil + film > fabric > maize straw + film. The treatment of film had significantly higher As concentration than the treatment of fabric and the treatment of maize straw + film.

There were no significant difference in Cd and Pb concentration among different treatments (Fig. 4.11). It is interesting to find that with the extended time, the Cd concentration in *P. vittata* increased very slightly, whereas Pb in *P. vittata* increased significantly.

The combination of plastic film and another layer of soil or maize straw can better improve soil temperature and survival rate. In terms of temperature boundary conditions for the growth of *P. vittata*, we propose that −2 °C is the lowest soil temperature that *P. vittata* can survive in.

Considering that phytoremediation greatly depends on the growth of plants, remediation efficiency is liable to changes in environmental conditions. Such uncertainty is the main issue scientists are trying to predict and control. The efficiency

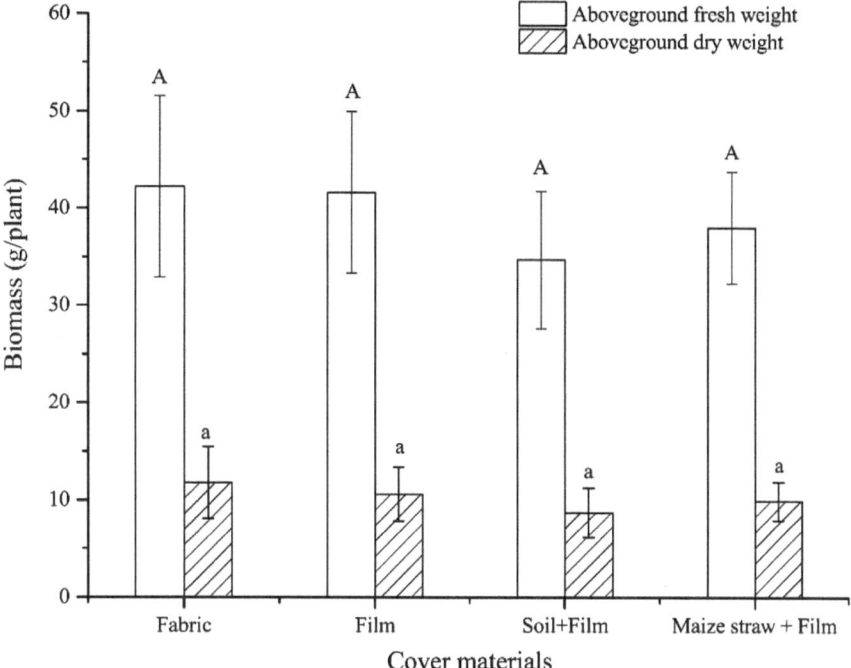

Fig. 4.9 Biomass with different cover materials

of the phytoremediation projects depend largely on the integrated ecosystem on the site (Barcelo and Poschenrieder 2011). Temperature is one of the main uncertainties that limits the application of As hyperaccumulator in cold regions (Wan et al. 2010). Through this experiment, the possibility of the application of *P. vittata* to As -contaminated soils in Northern China is confirmed.

The As concentration in soil after remediation apparently decreased. With all the collected samples showed an As concentration lower than the national standard. The distribution pattern was still the same as that before the remediation, indicating that *P. vittata* showed a strong As accumulating ability for the whole piece of land, not impacted by the heterogeneous As distribution. The Pb concentration was also decreased by the phytoremediation practice, but to a lesser extent than that of As.

Phytoextraction using hyperaccumulator *P. vittata* L. to extract As from soil has been applied to some contaminated areas in China, with highly-efficient results. However, cold winter weather is the main obstacle for this technology to be implemented in Northern China. Through the application of *P. vittata* in Jiyuan, Northern China, and the comparison of four cover materials, *P. vittata* has a strong potential to be used in cold regions. The phytoextraction project in Jiyuan, Henan Province, was the first one established in the monsoon climate of medium latitudes, with the previous projects all distributed in subtropical monsoon climate.

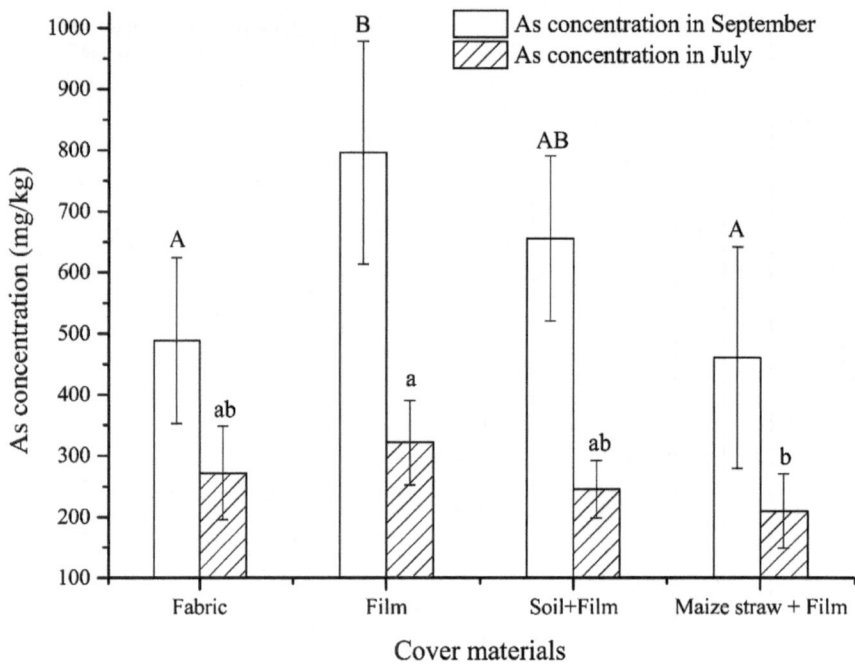

Fig. 4.10 As concentration of *P. vittata* covered by different materials

Also, a single layer of plastic film is the best cover material to obtain the highest As removal amount from each unit of soil. Although under natural conditions, *P. vittata* cannot survive in cold winter, the adoption of simple and cheap cover materials can efficiently and economically solve this problem. The cost for the cover material ranges from 5000 to 6000 $ per hectare soil, whereas the cost for the re-planting of *P. vittata* is as high as 30,000 $ per hectare soil. With the extended remediation period, the saved cost will further increase.

Through the above practices, phytoremediation is confirmed to be an economical, easy and green technology, which well fit for the current farmland contamination status. *P. vittata* showed potential to remove As from soil with As concentration ranging from 16.3 to 1184 mg kg^{-1} (Liao et al. 2004; Wan et al. 2016; Xie et al. 2010). The remediation efficiency showed variation. One reason for the different remediation efficiencies is the variation in As accumulating ability. The observed As concentration in *P. vittata* ranged from 409 to 1700 mg kg^{-1}. The remediation efficiency depended not only on the accumulating ability of As by *P. vittata*, but also on the environmental conditions, such as soil As concentration, As speciation and soil properties.

However, there are still a lot of questions waiting for further study.

1. The potential value of the harvested hyperaccumulator biomass. In France, the nickel (Ni) hyperaccumulator has been reused for the phytomining of Ni.

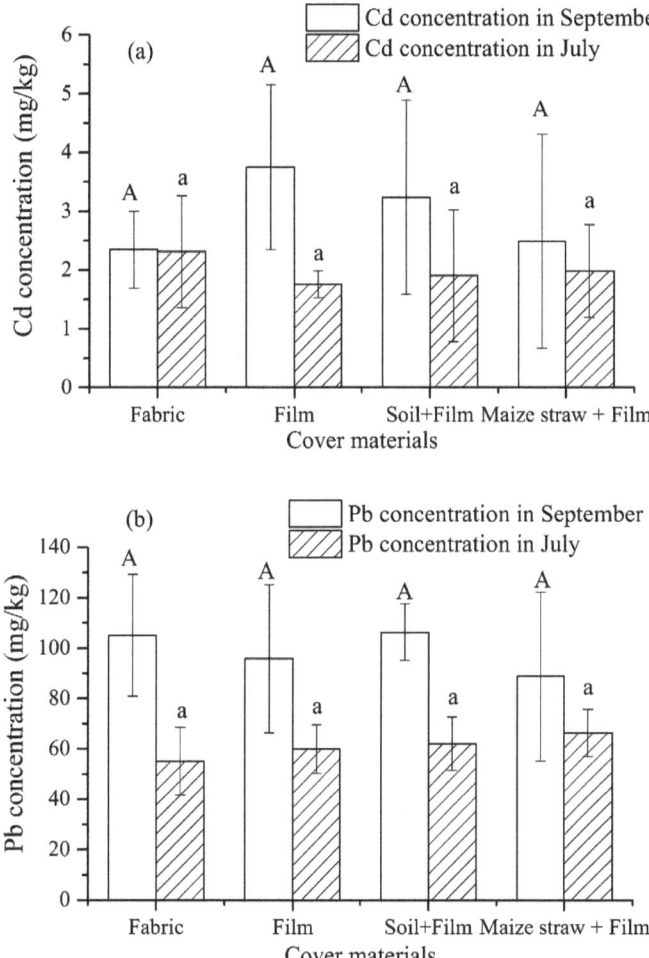

Fig. 4.11 Cd (**a**) and Pb (**b**) concentration of *P. vittata* covered by different materials

However, due to the low economical value of As, the potential reuse method for *P. vittata* still needs further investigation.

2. The control of weed. In the practice, it has been found that weed is a serious problem, which significantly increase the management cost. The development of an environment-safely herbicide for *P. vittata* is one of the next research targets.

3. The strengthening measures of phytoextraction. Compared to chemical or physical technologies, the time required by phytoextraction is long. Measures to promote the extraction of As from soil by *P. vittata* still requires further study.

References

Barcelo J, Poschenrieder C (2011) Hyperaccumulation of trace elements: from uptake and tolerance mechanisms to litter decomposition; selenium as an example. Plant Soil 341:31–35

Chen T, Lei M, Wan X, Yang J, Zhou X (2018a) Arsenic hyperaccumulator *Pteris vittata* L. and its application to the field. In: Luo Y, Tu C (eds) Twenty years of research and development on soil pollution and remediation in China. Singapore, Springer Singapore, pp 465–476

Chen T, Lei M, Wan X, Zhou X, Yang J, Guo G, Cai W (2018b) Element case studies: arsenic. In: Van der Ent A, Echevarria G, Baker AJM, Morel JL (eds) Agromining: farming for metals: extracting unconventional resources using plants. Springer International Publishing, Cham, pp 275–281

Chen TB, Fan ZL, Lei M, Huang ZC, Wei CY (2002a) Effect of phosphorus on arsenic accumulation in As-hyperaccumulator Pteris vittata L. and its implication. Chin Sci Bull 47:1876–1879

Chen TB, Liao XY, Huang ZC, Lei M, Li WX, Mo LY, An ZZ, Wei CY, Xiao XY, Xie H (2007) Phytoremediation of arsenic-contaminated soil in China. In: Willey N (ed) Phytoremediation. Humana Press, pp 393–404

Chen TB, Wei CY, Huang ZC, Huang QF, Lu QG, Fan ZL (2002b) Arsenic hyperaccumulator *Pteris vittata* L. and its arsenic accumulation. Chin Sci Bull 47:902–905

Ebbs S, Hatfield S, Nagarajan V, Blaylock M (2010) A comparison of the dietary arsenic exposures from ingestion of contaminated soil and hyperaccumulating *pteris* ferns used in a residential phytoremediation project. Int. J. Phytorem 12:121–132

Fu F, Song B, Zhong X, Wei S, Huang G (2015) Effects and risk assessment of heavy metals in sediments of Dahuanjiang River since Tailing Dam Break. Res Environ Sci 28:31–39

Gao F, Guo W, Wang J, Zhao X (2015) Historical record of trace elements input and risk in the shallow freshwater lake, North China. J Geochem Explor 155:26–32

Heimann L, Roelcke M, Hou Y, Ostermann A, Ma W, Nieder R (2015) Nutrients and pollutants in agricultural soils in the peri-urban region of Beijing: status and recommendations. Agr Ecosyst Environ 209:74–88

Li K, Liang T, Wang L, Yang Z (2015) Contamination and health risk assessment of heavy metals in road dust in Bayan Obo Mining Region in Inner Mongolia, North China. J Geog Sci 25:1439–1451

Liao X, Chen T, Xie H, Xiao X (2004) Effect of application of P fertilizer on efficiency of As removal form As contaminated soil using phytoremediation: field study. Acta Sci Circum 24:455–462

Long XX, Zhang YG, Jun D, Zhou QX (2009) Zinc, Cadmium and Lead accumulation and characteristics of rhizosphere microbial population associated with hyperaccumulator Sedum Alfredii Hance under natural conditions. Bull Environ Contam Toxicol 82:460–467

Ma LQ, Komar KM, Tu C, Zhang WH, Cai Y, Kennelley ED (2001) A fern that hyperaccumulates arsenic (vol 409, pg 579, 2001). Nature 411:438–U433

Tang Z, Zhang L, Huang Q, Yang Y, Nie Z, Cheng J, Yang J, Wang Y, Chai M (2015) Contamination and risk of heavy metals in soils and sediments from a typical plastic waste recycling area in North China. Ecotoxicol Environ Saf 122:343–351

Wan X, Lei M, Chen T (2016) Cost–benefit calculation of phytoremediation technology for heavy-metal-contaminated soil. Sci Total Environ 563–564:796–802

Wan XM, Lei M, Chen TB, Zhou GD, Yang J, Zhou XY, Zhang X, Xu RX (2014) Phytoremediation potential of *Pteris vittata* L. under the combined contamination of As and Pb: beneficial interaction between As and Pb. Environ Sci Pollut Res 21:325–336

Wan XM, Lei M, Huang ZC, Chen TB, Liu YR (2010) Sexual propagation of *Pteris vittata* L. Influenced by pH, Calcium, and temperature. Int J Phytorem 12:85–95

Wang S, Wei S, Ji D, Bai J (2015) Co-Planting Cd contaminated field using hyperaccumulator *Solanum nigrum* L. Through interplant with low accumulation Welsh Onion. Int J Phytorem 17:879–884

Xie J, Lei M, Chen T, Li X, Gu M, Liu X (2010) Phytoremediation of soil co-contaminated with arsenic, lead, zinc and copper using *Pteris vittata* L.: a field study (in Chinese, abstract in English). Acta Sci Circum 30:165–171

Yang XE, Ye HB, Long XX, He B, He ZL, Stoffella PJ, Calvert DV (2004) Uptake and accumulation of cadmium and zinc by *Sedum alfredii* Hance at different Cd/Zn supply levels. J Plant Nutr 27:1963–1977

Zhang Y, Wan X, Lei M (2016) Application of arsenic hyperaccumulator *Pteris vittata* L. to contaminated soil in Northern China. J Geochem Explor

Chapter 5
Evaluation of Phytoremediation Efficiency: Field Experiences

Abstract The arsenic (As) phytoremediation technology, based on the As hyper-accumulating ability of *Pteris vittata*, has been applied to more than 20 phytoremediation projects, achieving different levels of remediation efficiencies. However, currently, there are no commonly accepted evaluation methods of phytoremediation efficiency. The decrease in soil As concentration, the accumulation of As in hyperaccumulating plants, and the bioavailability of As in soil are factors to be considered. This chapter briefly introduced the application of these three indexes to the evaluation of phytoremediation efficiency. Recently, it has been further proposed that the remediation of agricultural soil should also take sustainability into consideration. Therefore, further study is needed to establish a streamline for a comprehensive evaluation of phytoremediation efficiency.

Keywords Arsenic · Distribution · Efficiency · Evaluation · Sustainability

The arsenic (As) phytoremediation technology utilizing the hyperaccumulating fern *Pteris vittata* L., including phytoextraction and intercropping, has been applied to 400 ha of soil in 12 sites with a removal rate ranging from 10% to 17% per year (Chen et al. 2018a, b; da Silva et al. 2018). During these phytoremediation practice, it has been found that the method for evaluation of phytoremediation efficiency needs to be unified. Otherwise, it is hard to compare different phytoremediation projects. The decrease in soil As content, the accumulated amount of As in the hyperaccumulator harvests, and the As concentration in the agricultural products after remediation are indexes that have been used in the previous chapters.

In this chapter, evaluation methods using the decrease in soil As concentration, and the accumulation of As in hyperaccumulating plants were used. Besides, the discrepancy between these two methods and possible reasons were analyzed.

© The Author(s) 2020
T. Chen et al., *Phytoremediation of Arsenic Contaminated Sites in China*,
SpringerBriefs in Environmental Science,
https://doi.org/10.1007/978-981-15-7820-5_5

5.1 Evaluation Based on the Decrease in Soil as Concentration

In one of the early phytoremediation projects, a five-year phytoremediation project was established on a piece of farmland contaminated by wastewater and slag discharge from a nearby As products factory. The As hyperaccumulator was planted in April 2002 and harvested in December each year from 2002 to 2007. The phytoremediation efficiency was evaluated using the decrease in soil As concentrations (Liao et al. 2004; Liu et al. 2005). In the first year, after 7 months of phytoremediation, soil As concentrations were significantly reduced, with the highest removal rate reaching 7.84% in the treatment applied with 369 kg P hm^{-2}. From 2002 to 2007, apparent decrease was also found in the concentration of the bioavailable As (extracted by $NaHCO_3$) in the heavily contaminated areas, but not found in the moderately or lightly contaminated soils (Table 5.1). However, it should be noted that the $NaHCO_3$-extractable As concentration in soil showed large deviation, indicating a heterogeneous distribution, despite that the soil has been plowed before the experiment.

With the undergoing of the remediation project, the heterogeneous distribution of As in soil becomes more and more apparent. In the second year, soil samples were collected from non-rhizosphere and rhizosphere soil from heavily contaminated soil, moderately contaminated area and lightly contaminated soil, respectively. Results indicated significant difference in As concentration not only among different contaminated areas but also between non-rhizosphere and rhizosphere soil (Fig. 5.1).

In the slightly contaminated soil, the concentration of As in rhizosphere soil was higher than that in non-rhizosphere soil, which might be related to the application of phosphorus fertilizer as a strengthening measure. In moderately and heavily contaminated soil, the As concentration in rhizosphere soil was lower than that in non-rhizosphere soil, which might be related to the strong absorption of As by *P. vittata*.

The enrichment of As in the rhizosphere soil was not uniformly observed for all the sampling locations, which further indicated the heterogeneously spatial distribution of As in soil.

Table 5.1 Change in the $NaHCO_3$-extracted concentration of As in soil

Year	Extracted As (mg kg^{-1})		
	Heavily contaminated	Moderately contaminated	Lightly contaminated
2002	7.16 ± 9.41a	1.11 ± 0.32a	0.46 ± 0.25a
2004	5.92 ± 7.53a	0.95 ± 0.43a	0.51 ± 0.26a
2005	7.19 ± 10.14a	1.21 ± 0.42a	0.41 ± 0.27a
2006	6.16 ± 7.69a	1.30 ± 0.66a	0.49 ± 0.26a
2007	6.40 ± 8.45a	1.02 ± 0.49a	0.63 ± 0.30a

Fig. 5.1 As concentrations in non-rhizosphere and rhizosphere soils growing with *P. vittata* L

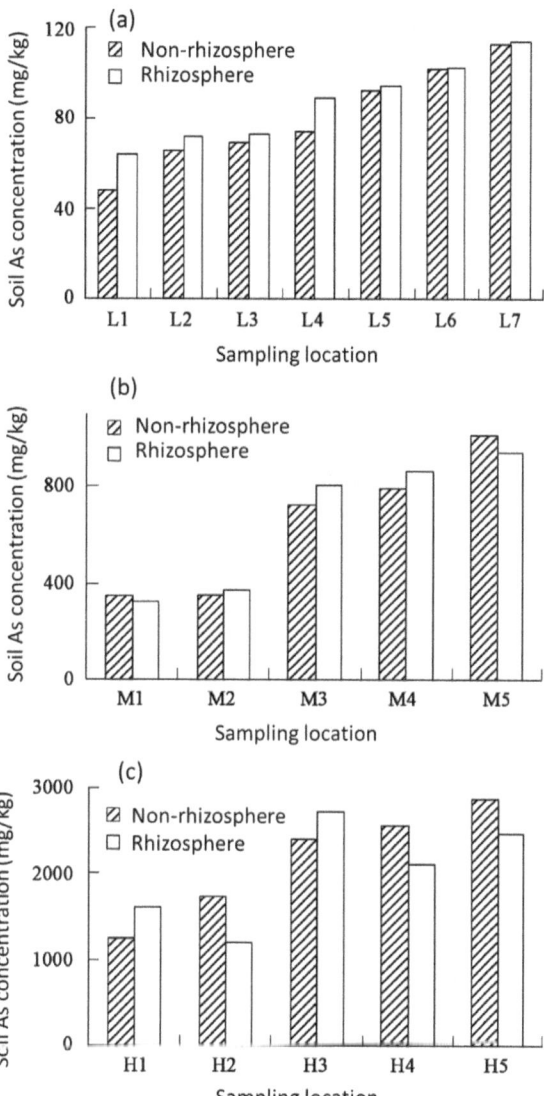

Therefore, the later evaluation of phytoremediation efficiency through comparing the soil As concentration before and after remediation not only considers the decrease in single point concentration of As, but paid more attention to the change in the spatial distribution of As.

A two-year field experiment evaluated the phytoremediation efficiency using the hyperaccumulator *P. vittata* on an As contaminated site by comparing the As concentration (Fig. 5.2) and As distribution pattern (Fig. 5.3).

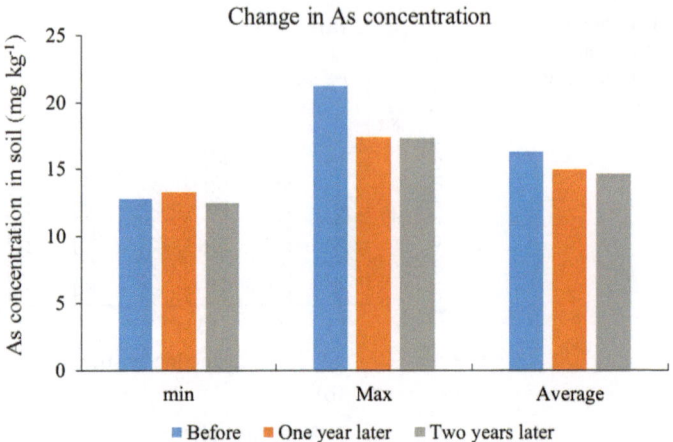

Fig. 5.2 The concentration of As in soil before and after remediation

Despite the fact that As concentration in the soil was lower than the value that recommended by China's Environmental Quality Standard for Soils (25 mg kg^{-1}), the As concentrations in 7% of the wheat and maize samples exceed the national standard for food (0.5 mg kg^{-1}). This indicated a potential threat of As on the human health. The reason for the exceptionally high concentration of As in wheat when soil As concentration meet the national standards may have resulted from the high heterogeneity of As in soil.

High concentrations of As were mainly distributed in the east of the remediation site (Fig. 5.3a). The As concentrations in the east of the remediation site was as high as 24 mg kg^{-1}. While the rest part showed much lower concentration of As in soil.

After one year of planting the As hyperaccumulator *P. vittata*, the average As concentration in soil decreased to 14.99 mg kg^{-1}, ranging from 13.34 to 17.45 mg kg^{-1}. After two-year remediation, the average As concentration in soil decreased to 14.58 mg kg^{-1}, ranging from 12.47 to 17.32 mg kg^{-1}. The distribution of As also indicated apparent decrease in As concentration (Fig. 5.3b). And it is note worthy that the As concentration in the east part where higher concentration of As was found showed apparent decrease after remediation.

After one-year remediation, the removal rate of As in soil was ~7.87%, which decreased to ~2.74% in the second year. The removal rate of As after two-year remediation was 10.39%.

And it is apparent that using the change in spatial distribution of As instead of average concentration of As is a more comprehensive way to evaluate the remediation efficiency. It also needs to mention that different interpolation methods could result in varied distribution pattern of elements, especially for the areas with exceptionally high or low values. Increasing the sampling density might improve the interpolation accuracy.

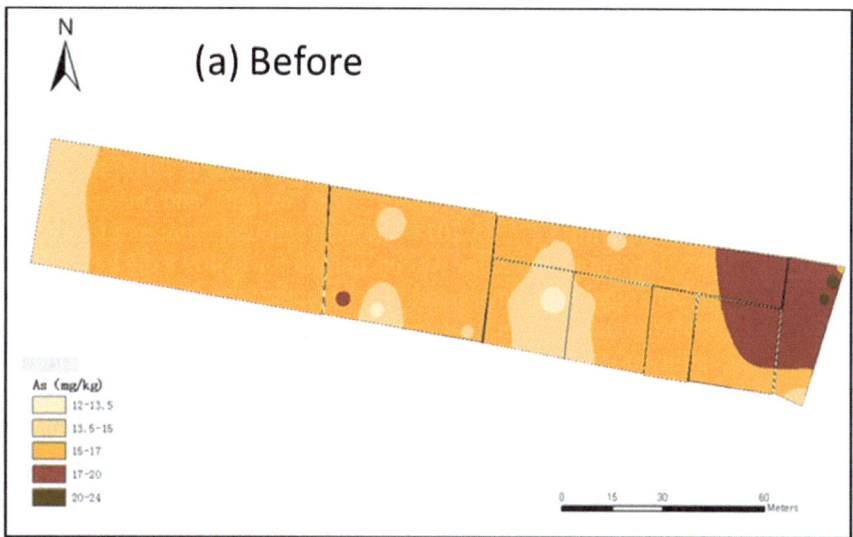

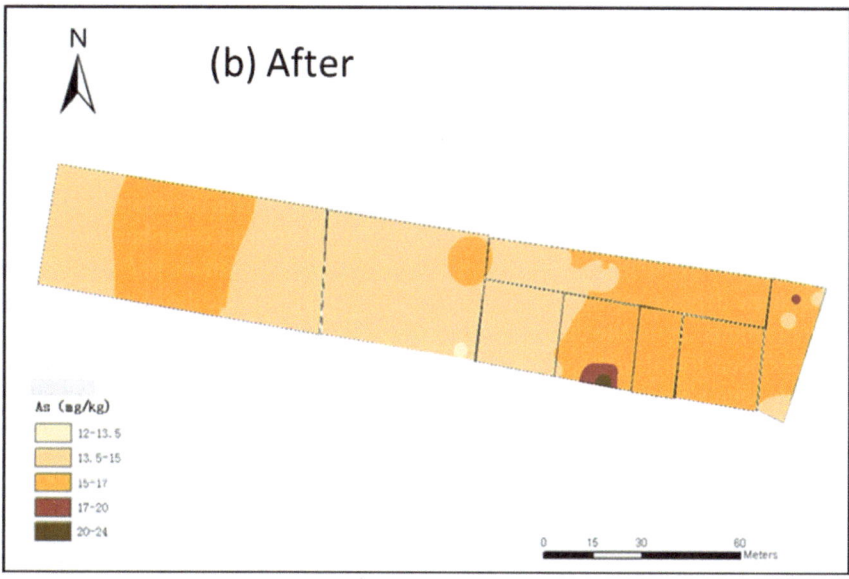

Fig. 5.3 The distribution of As before (**a**) and after (**b**) remediation

5.2 Evaluation Based on Arsenic Amount Accumulated in Hyperaccumulating Plant

Evaluation based on the As amount accumulated in the hyperaccumulating plant considers both the biomass and the As concentration in plants. In 2013, the above-ground biomass of *P. vittata* was 7000 kg hm^{-2} (dry weight), and the average As concentration was 419.6 mg kg^{-1}. In 2014, the aboveground biomass of *P. vittata* was 6909 kg hm^{-2} (dry weight), and the As concentration was 409.8 mg kg^{-1}. There is no significant difference in biomass and As concentration of *P. vittata* between 2013 and 2014. The bioaccumulation factor of *P. vittata* reached 28, indicating high accumulating ability of *P. vittata*. Results indicate that *P. vittata* grew well on this remediation site, and continuously took up As from soil.

After one-year remediation, the removal rate of As in soil was ~7.87%, which decreased to ~2.74% in the second year. The removal rate of As after two-year remediation was 10.39%.

Comparing the decrease in soil as concentration and the extracted As by *P. vittata*, it seems that although As was continuously taken up from soil from the results of As concentration in plants, As in soil did not show apparent decrease from the results of As concentration in soil. By multiplying biomass and As concentration, it was calculated that 5768 g As was removed from one ha soil after two-year remediation. The calculated phytoextraction ratio was 16.09%. Where is the gap between 10.39% (calculated from the decrease in soil As concentration) and 16.09% (calculated from the accumulation of As in plants)?

Possible reasons were analyzed. This remediation site was in the downwind direction of a Pb smelter. Therefore, atmospheric deposition might contribute to the gap between the decrease in soil As content and the accumulation of As in *P. vittata*. The atmospheric deposition was monitored every month, to evaluate the accumulation of As in the remediation site due to atmospheric deposition.

It has been found that compared to dusts collected at the control point with no pollution source around, the dusts collected at the remediation site contained a considerable amount of As (Fig. 5.4). In the control points, the deposited As was lower than 1 kg km^{-2} in 30d, whereas in the remediation site, the deposited As reached 20 kg km^{-2} in 30d. Results indicated that there existed apparent atmospheric emission in this area, which may have led to the insignificant difference in soil As concentration despite the continuous extraction of As from soil by *P. vittata*.

Taking into consideration the effects of atmospheric deposition, 0.6 mg As kg^{-1} soil was added each year, which needs to be corrected. After deductions of increase in soil As due to atmospheric deposition, the corrected removal rate of As in soil reached 16.57% after two-year remediation. This was basically in accordance with the efficiency calculated according to the accumulated As in *P. vittata*.

With the development of phytoextraction technology, the proper method to evaluate the efficiency of phytoextraction becomes increasingly important. Using removal rate of As from soil is the most common method. However, as the above case indicated, due to the large differences in soil As concentration, using decrease in

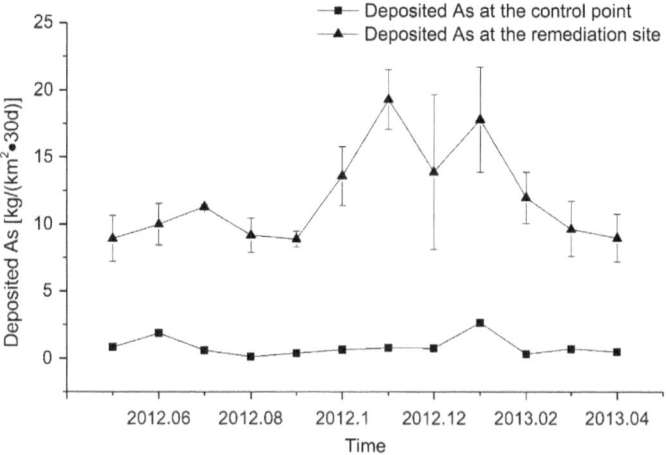

Fig. 5.4 Atmospheric deposition of As in the control and remediation sites

soil concentration may be not comprehensive enough (Niazi et al. 2012). Therefore, both the decrease in soil HMs concentration and the increase in hyperaccumulator concentration were chosen as essential indexes.

Further, the complete input and output cycle of HMs in the ecosystem should be considered. The input of HMs into soil includes the atmospheric deposition, irrigation and fertilization, canopy leaching, hyperaccumulator litters (Barcelo and Poschenrieder 2011). The output of HMs from soil includes hyperaccumulator extraction, surface runoff, soil infiltration and plant surface evaporation. In the remediation site described in the current study, atmospheric deposition was considered, being one of the most important inputs. Only after incorporating this input, the phytoextraction efficiency calculated from the aspects of hyperaccumulator removal and soil decrease matched.

Considering the two methods mentioned here, it seems that change in soil As concentration is a more accurate and direct way to evaluate the overall change of soil environmental quality, while the removed As by plants can be used as a secondary index reflecting the efficiency of phytoremediation.

5.3 The Change in the Bioavailability of As

With the update of the mainstream environmental management policy in recent five years, the phytoremediation technology gradually also paid attention to the bioavailability of As instead of the total concentration of As. This was especially useful in one of the phytoremediation technologies: intercropping. Intercropping can also be regarded as a safety utilization method for slightly contaminated soil.

Intercropping is one of the traditional planting system, which can produce more products on a certain piece of land (Zhang and Li 2003). Recently, intercropping has been used to manage slightly As contaminated soil, which can decrease As concentration in soil, and at the same time produce agricultural products which satisfy the national standards (Wang et al. 2015). The intercropping technology has been applied to several contaminated farmland in south China.

Since one of the aims of intercropping is to produce safe agricultural products, the evaluation of the phytoavailability of As in soil becomes more important. A tank experiments was conducted to compare the total concentration, available concentration and spatial distribution of As in soil from the treatments of no plants, monoculture of cash crop *Morus alba*, monoculture of *P. vittata*, and the intercropping of *M. alba* and *P. vittata*.

Results indicated that the concentration of As in the soil solution was in the order of no plant > monoculture of *M. alba* > monoculture of *P. vittata* ≈ intercropping of *M. alba* and *P. vittata* (Fig. 5.5). The As concentration of soil solution in the monoculture system of *P. vittata* and in the intercropping system of *M. alba* and *P. vittata* was the lowest among these four treatments, which followed a similar decrease trend from ~100 μg L^{-1} to ~45 μg L^{-1} after seven-week cultivation. In the treatment with no plant growing, the As concentration in the soil solution decreased from 133 μg L^{-1} to 84 μg L^{-1} after seven-week cultivation. The As concentration in the monoculture of *M. alba* decreased from 113.6 μg L^{-1} at the end of the first week to 96.2 μg L^{-1} at the end of the experiment.

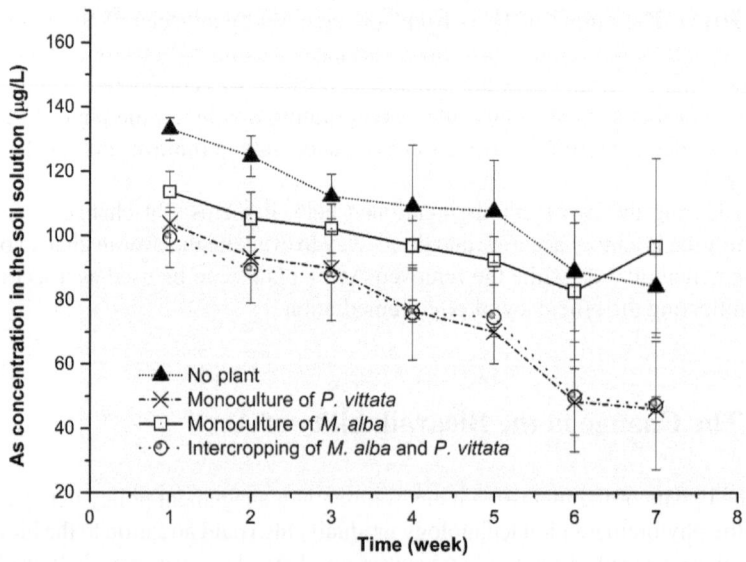

Fig. 5.5 As concentration in the soil solution in monoculture and intercropping systems of *M. alba* and *P. vittata* (Wan et al. 2017)

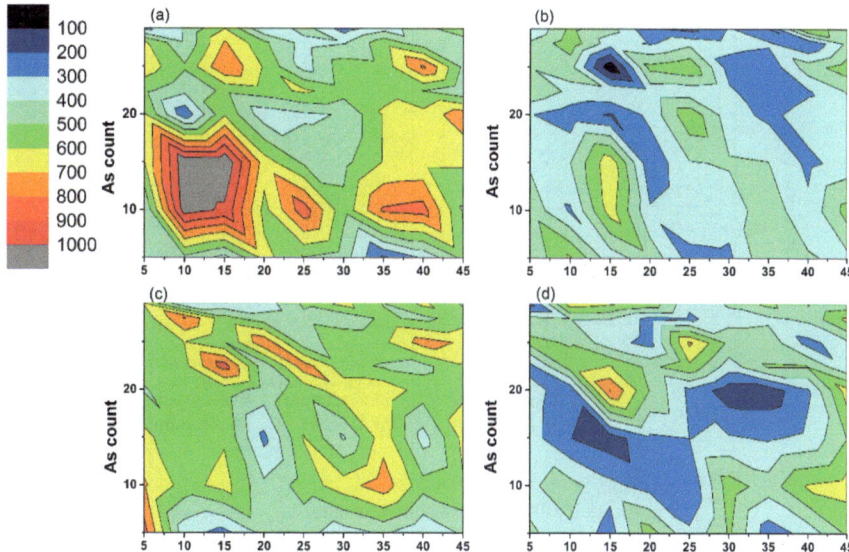

Fig. 5.6 As distribution in the rhizospheres of monoculture and intercropping systems of *M. alba* and *P. vittata* (a. no plant; b. monoculture of *P. vittata*; c. monoculture of *M. alba*; and d. intercropping of *P. vittata* and *M. alba*)

Spatial distribution of As directly indicated the change in As concentration in soil. In the intercropping system, As concentration in the rhizosphere soil was significantly lower than that in the monoculture system of *M. alba* (Fig. 5.6). In the treatment without plants, As accumulated at the bottom left of the tank, which may be related to the location of the valve of the tank. In the monoculture system of *M. alba*, As accumulated at the rhizosphere of *M. alba*. By contrast, in the monoculture system of *P. vittata*, As concentration was apparently decreased at the rhizosphere layer, indicating the depletion of As by the roots of *P. vittata*. Similarly, As in the rhizosphere of *P. vittata* was also depleted in the rhizosphere of *P. vittata*.

Compared with the monoculture of *M. alba*, intercropping of *M. alba* with *P. vittata* decreased the leaf As concentration of *M. alba* by 41.5%. Intercropping also decreased the mobility of As in the rhizosphere. This may release the worry about the increased risk of As migration in the intercropping system of *M. alba* and *P. vittata*.

In accordance with the soil solution result, sequential extraction experiment found that the fraction of As associated with amorphous (hydr)oxides was significantly lower in the intercropping treatment than the monoculture treatments, accompanied by an increase in the fraction of As associated with crystalline (hydr)oxides (Fig. 5.7). According to Wenzel et al. (2001), non-specifically sored As (fraction 1) can represent solute As, while the specifically sorbed As (fraction 2), amorphous (hydr)oxides associated As (fraction 3) and crystalline (hydr)oxides associated As (fraction 4) indicate potential lability of As from different solid phases as a result of alteration in soil, such as pH or Eh. The residue (fraction 5) is considered as a comparatively

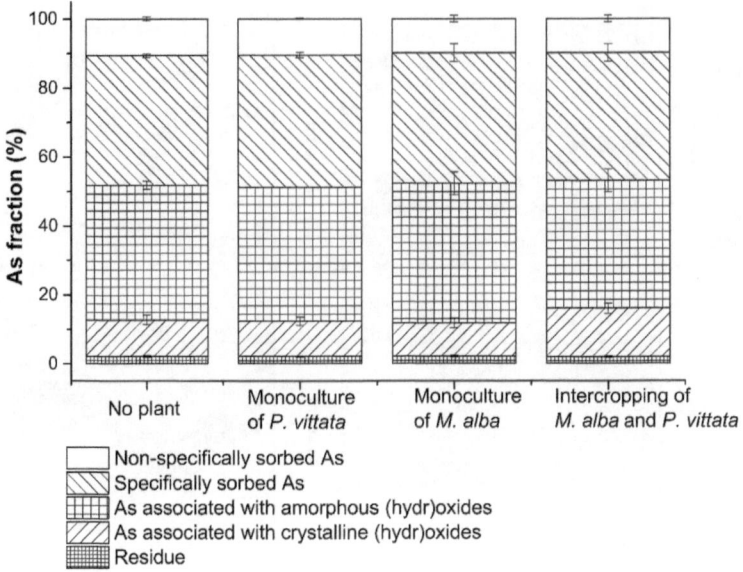

Fig. 5.7 The fraction of As rhizosphere soil in the monoculture and intercropping systems of *M. alba* and *P. vittata*

stable fraction. From fraction 1 to fraction 5, the mobility of As decreased. Therefore, the transfer of As from fraction 3 to fraction 4 observed in the current study indicated that intercropping helped immobilize As in the soil.

Therefore, when evaluating the intercropping efficiency, not only the decrease in soil total As concentration needs to be considered, but also the availability of As needs to be considered.

5.4 Future Development of the Evaluation Method for Phytoremediation Efficiency

Recently, a more comprehensive evaluation frame of remediation efficiency from a sustainability view has been proposed (Hou et al. 2018; O'Connor et al. 2019). Green and sustainable remediation is a global trend. Agricultural land remediation also needs to consider social and economic considerations. The ultimate goal of any remediation process should be not only to remove the contaminant(s) from the polluted site but to restore the continued capacity of a soil to perform or function according to its potential (i.e., its health) as well (Hernandez-Allica et al. 2006). This implies that it is necessary to carry out further research on the appropriate evaluation method for the soil remediation efficiency, and this will be out next research focus.

Besides the three index categories usually used during the sustainability assessment for remediation at industrial sites (social, economic, and environmental aspects), the evaluation of agricultural land remediation efficiency brings in another category (agricultural), which originates from agricultural soil quality assessment (Hou et al. 2018). These four categories are further divided into 11 subcategories, and in total 32 quantifiable indicators. And during the calculation, these 32 indicators were given the same weight. This assessment work indicated that most of the reviewed agricultural soil remediation projects need to improve the social and agricultural efficiency, and the long-term efficacy of the remediation technologies. Many of these concerns are not considered during traditional human health risk based decision making processes. This frame is currently only used during the decision-making processes, which should be used for the retrospective evaluation of remediation efficiency in the future.

References

Barcelo J, Poschenrieder C (2011) Hyperaccumulation of trace elements: from uptake and tolerance mechanisms to litter decomposition; selenium as an example. Plant Soil 341:31–35

Chen T, Lei M, Wan X, Yang J, Zhou X (2018a) Arsenic hyperaccumulator *Pteris vittata* L. and Its application to the field. In: Luo Y, Tu C (eds) Twenty years of research and development on soil pollution and remediation in China. Singapore, Springer Singapore, pp 465–476

Chen T, Lei M, Wan X, Zhou X, Yang J, Guo G, Cai W (2018b) Element case studies: Arsenic. In: Van der Ent A, Echevarria G, Baker AJM, Morel JL (eds) Agromining: farming for metals: extracting unconventional resources using plants. Springer International Publishing, Cham, pp 275–281

da Silva EB, Lessl JT, Wilkie AC, Liu X, Liu Y, Ma LQ (2018) Arsenic removal by As-hyperaccumulator *Pteris vittata* from two contaminated soils: a 5-year study. Chemosphere 206:736–741

Hernandez-Allica J, Becerril JM, Zarate O, Garbisu C (2006) Assessment of the efficiency of a metal phytoextraction process with biological indicators of soil health. Plant Soil 281:147–158

Hou DY, Ding ZY, Li GH, Wu LH, Hu PJ, Guo GL, Wang XR, Ma Y, O'Connor D, Wang XH (2018) A sustainability assessment framework for agricultural land remediation in China. Land Degrad Dev 29:1005–1018

Liao X, Chen T, Xie II, Xiao X (2004) Effect of application of P fertilizer on efficiency of As removal form As contaminated soil using phytoremediation: field study (in Chinese, and abstract in English). Acta Sci Circum 24:455–462

Liu Y-R, Chen T-B, Huang Z-C, Liao X-Y (2005) As-hyperaccumulation of *Pteris vittata* L. as influenced by as concentrations in soils of contaminated fields. Huan jing ke xue = Huanjing kexue 26, 181–186

Niazi NK, Singh B, Van Zwieten L, Kachenko AG (2012) Phytoremediation of an arsenic-contaminated site using *Pteris vittata* L. and Pityrogramma calomelanos var. austroamericana: a long-term study. Environ Sci Pollut Res 19:3506–3515

O'Connor D, Zheng XD, Hou DY, Shen ZT, Li GH, Miao GF, O'Connell S, Guo M (2019) Phytoremediation: climate change resilience and sustainability assessment at a coastal brownfield redevelopment. Environ Int 130

Wan X, Lei M, Chen T, Yang J (2017) Intercropped *Pteris vittata* L. and *Morus alba* L. presents a safe utilization mode for arsenic-contaminated soil. Sci Total Environ 579:1467–1475

Wang SQ, Wei SH, Ji DD, Bai JY (2015) Co-planting Cd contaminated field using hyperaccumulator *Solanum nigrum* L. Through interplant with low accumulation Welsh Onion. Int J Phytorem 17:879–884

Wenzel WW, Kirchbaumer N, Prohaska T, Stingeder G, Lombi E, Adriano DC (2001) Arsenic fractionation in soils using an improved sequential extraction procedure. Anal Chim Acta 436:309–323

Zhang FS, Li L (2003) Using competitive and facilitative interactions in intercropping systems enhances crop productivity and nutrient-use efficiency. Plant Soil 248:305–312